Macdonald

9-50

A20

CLINICAL RADIOBIOLOGY

Clinical Radiobiology

W. Duncan F.R.C.P.E., F.R.C.R.

Professor of Radiotherapy, University of Edinburgh;
Honorary Consultant Radiotherapist, Royal Infirmary and
Western General Hospital, Edinburgh.

A. H. W. Nias M.A., D.M. (Oxford) D.M.R.T.

Consultant Medical Radiation Biologist, Glasgow Institute of
Radiotherapeuties and Oncology, Belvidere Hospital,
Glasgow. Honorary Clinical Lecturer, University of Glasgow.

CHURCHILL LIVINGSTONE
Edinburgh London and New York 1977

CHURCHILL LIVINGSTONE
Medical Division of Longman Group Limited

Distributed in the United States of America by Longman Inc.,
19 West 44th Street, New York, N.Y. 10036 and by associated
companies, branches and representatives throughout the
world.

ISBN 0 443 01147 8

Library of Congress Cataloging in Publication Data

Duncan, William.
    Clinical radiobiology.

    1. Radiobiology. 2. Radiotherapy.
I. Nias, A. H. W., joint author. II. Title.
[DNLM: 1. Radiobiology. WN610 D912c]
RM849.D8   616.07'57   76-28527

Printed in Great Britain by
Butler & Tanner Ltd
Frome and London

PREFACE

There is no doubt that it is now essential for radiotherapists to have a sound basic knowledge of radiobiology. This has recently been recognised by the Royal College of Radiologists, London, which now requires that radiotherapists in training have formal instruction in radiobiology and the subject is an important part of the First Examination for their Fellowship.

The material in this book is designed to cover the syllabus in radiobiology recommended by the Royal College of Radiologists and it has been written with a deliberately clinical bias. As a result, many of the more fundamental principles of radiobiology have received comparatively little attention in this book, while many aspects of applied radiobiology of less scientific importance have been emphasised because they have direct relevance to clinical practice.

Although the book is written primarily for the clinician who wishes to have some understanding of the basic biological principles of radiotherapy, it is hoped that it will prove of value and interest to others such as physicists, biologists and radiographers who are involved in the management of patients in Departments of Radiotherapy and Oncology.

The authors wish particularly to thank Dr. J. F. Fowler of the Gray Research Laboratory of the Cancer Research Campaign, Mount Vernon Hospital, London who read the script and made many constructive suggestions.

We also have to express our gratitude to our secretaries, Mrs L Manson and Mrs G Logan who undertook the typing and secretarial work in the preparation of the manuscript. We are particularly indebted to Mr. J. Pizer who has redrawn all the illustrations and to Mr R. Robertson who was responsible for the photographic and art work. We would express our thanks to them for their advice and professional services.

Finally, we take pleasure in acknowledging our thanks to the staff of Churchill Livingstone for their advice, patience and understanding during the writing of the book.

W.D.
A.H.W.N.

Edinburgh, 1977

*At the time this book is being printed, S.I. units are being introduced into the scientific community. Despite this, we have decided to use traditional units because most readers will be more familiar with them. Of the radiological units, that for absorbed dose, the rad, is the one most commonly found in this book, and readers may, simply by dividing by 100 convert to the S.I. unit, the gray (Gy). The conversion of the unit for radioactivity is less simple; (e.g. 1 megabecquerel = 27.03 microcuries), but this unit does not occur very often in the book.

CONTENTS

1. Introduction

The practice of radiotherapy is founded almost entirely on the experience and carefully documented clinical observations of skilled and studious radiologists over the last 50 years. Only recently has its basic science, radiobiology, begun to influence the understanding and development of the clinical applications of radiations used in cancer therapy. The reasons for this apparently late influence are of interest and are best explained by relating a short account of the history of radiation research.

The origin of the subject is, of course, to be found in the discovery of X-rays by Professor Wilhelm Conrad Roentgen in the University of Würzburg. The announcement on December 28th 1895 of this new type of radiation attracted immediate interest throughout the world. In the next year Henri Becquerel, a French physicist, reported to the Academy of Sciences in Paris of the 'emanation' from uranium, and soon afterwards Marie and Pierre Curie were able to isolate radium and introduced the concept of 'radioactivity' to the world of science.

In 1934 the first artificial radioactivity was produced by Frederic Joliot and his wife Irene, daughter of Marie Curie. Progress in atomic physics was rapid thereafter and in 1939 the 'atomic age' was born with the demonstration of nuclear fission by Hahn and Strassman in Germany.

The early years of the development of radiotherapy were dominated by the improvements made in radiation physics, engineering and technology. It was obviously essential to have safe and reliable X-ray machines with energies capable of sufficient penetration in tissues to treat deep-seated tumours. Methods of accurate radiation dose measurement had to be evolved to ensure the consistent exposure of the area to be treated.

A major advance was achieved with the manufacture of the Coolidge hot filament tube in America in 1913. This generator was able to provide for the first time a controlled and reliable output of X-radiation and was to be the model for future ortho-voltage X-ray generators which became available in the early 1920's. There followed in 1932 the production of the first Cyclotron, the Van de Graaf high-voltage generator and in 1940 the development of the Linear Accelerator for the production of megavoltage X-radiation and electron beams.

For many years the measurement of dose depended on chemical dosimetry, estimated by a change of colour as in the *Sabouraud et Noire* pastille unit. The system was difficult to standardize and many biological methods were used to supplement the chemical estimation of radiation exposure. A common practice was to determine the 'skin erythema dose' which produced a reddening of the skin in one or two weeks after irradiation. This too was wholly unsatisfactory, but it was not until 1928 that the Roentgen (R) unit was accepted internationally as the measure of ionisation in the air. Further improvement in clinical dosimetry was later realized with the introduction in 1954 of the rad, the unit of 'radiation absorbed dose'.

In spite of the technical difficulties and limitations of the primitive X-ray generators many forms of experimental treatment were tried at the earliest opportunity. Within four months of Roentgen's discovery Dr Daniels in the United States reported in April 1896 the loss of hair in one of his colleagues following irradiation. It is usually claimed that as a result of this report the first rational application of X-ray therapy was given by Dr Leopold Freund of Vienna, who successfully treated a benign hairy naevus in 1897. It may be that the first use of radiation in cancer therapy took place earlier in January 1896 when E. H. Grubbe, a physicist at the Hahneman Medical College in Chicago, claimed to have treated a patient with breast cancer referred to him by Dr Ludram (Hodges, 1964). Reports of other miscellaneous applications were soon published, but it was not until 1922 that radiotherapy could be recognized as a defined clinical discipline with a significant contribution to make in cancer control. It was at this time that Regaud, Coutard and Hautant (1922) reported the results of a series of patients with cancers of the larynx before the International Congress of Otology in Paris.

During the early days of experimental research and clinical practice employing X-rays, some unfortunate results of over-exposure were observed, especially on the skin of radiologists and X-ray workers. These and other pathological changes that had been reported encouraged extensive research on the pathology of radiation injury in all the organs and tissues. The work was essentially descriptive and qualitative and several important original reports and comprehensive monographs on radiation pathology were produced at this time, (Warren, 1928; Desjardins, 1931).

Soon after reliable X-ray equipment became available in the 1920's radiation research workers began to examine the nature of the biophysical events produced in tissues by X-rays. Initially the main interest lay in the effects of direct ionisation in the track of the radiation. It soon became clear that ionisation produced indirectly by the radiation as a result of free radical formation, principally by the radiolysis of water, was also of great importance. In this period many important basic studies were conducted by F. G. Spear and D. E. Lea in England and Timofeaff-Rissovsky and K. G. Zimmer in Germany. Their research culminated in the development of the 'target theory' of radiation action which was to lead to the quantitative evaluation of radiation effects on cells and important advances in radiobiological research.

It was not until 1927 that the first quantitative biological studies in radiation research were reported by H. J. Muller, who measured the rates of mutation after radiation in the *Drosophila* fruit fly. The dramatic breakthrough in radiobiology which has encouraged a most productive era of research came in 1950 when *in vitro* cell culture techniques were perfected and the era of quantitative cellular radiobiology was entered. At the same time increased concern and interest in the hazards of atomic radiation gave further impetus to the expansion of radiobiological research. There were now available reliable radiation sources, consistent dosimetry and a whole new range of quantitative techniques in cell biology. In 1953 Howard and Pelc described the phase progression of the cell cycle and opened a new field of cellular biology. In 1956 Puck and Marcus first reported their results of clone counting experiments following irradiation of mammalian cells, demonstrating the relationship of dose to cell survival.

These experiments were followed by a series of important discoveries which introduced radiobiologists and their clinical colleagues to a new language to describe the response of cells and tissues to radiation and to new concepts about the biological basis of radiotherapy. In this period it should be recognized that the contribution of radiobiology to clinical radiotherapy was mainly conceptual and few practical applications based on laboratory findings were considered suitable for evaluation in clinical trials.

The serious effects of cancer are caused mainly by the progressive accumulation of the abnormal cells in the patient. The object of radiotherapy is to sterilize the tumour or to cause *loss of reproductive integrity* of the cancer cells. This term implies that the cell has lost its capacity for unlimited proliferation although the injured cell may retain the ability to go through several cell divisions before finally dying. During this time the cell will be apparently morphologically intact and differentiated cells may function and continue, for example, to produce in the case of a thyroid cell its complex hormone, just as it did before it was lethally irradiated. The mammalian cell will normally die after a lethal dose of radiation only in mitosis. For this reason rapidly dividing cells will show evidence of radiation damage more quickly than cells with long cell cycle times. The rate of response of cells to radiation injury is, therefore, related to their rate of turnover. It should be clear that it is not cells themselves, but the processes of cell division that are sensitive to the effects of radiation.

In 1959, Elkind and Sutton had elegantly demonstrated that mammalian cells possess the ability of *recovery from sub-lethal radiation damage* and that when recovery is complete the cells respond to subsequent doses of radiation as if they had never been irradiated. This is undoubtedly one of the most important fundamental processes in the response of cells to radiation and was of immediate relevance to radiotherapy. The 'wasted' radiation used in producing repairable damage explains in the main the need to increase the dose of radiation when many dose fractions are given. Also differences in degree of 'recovery' between some tumour cells and some normal tissue cells may be the single most important factor in explaining the selective eradication of the cancer cell population compared to normal tissues in successful radiotherapy.

Another phenomenon which is of major importance in radiobiology is the 'oxygen effect' first described in detail by L. H. Gray and his colleagues in 1953. Oxygen is the most powerful radiation sensitizer that has so far been described. The demonstration that many tumours contain a proportion of anoxic cells has led to the conclusion that these cells by virtue of their anoxia are relatively radio-resistant and may be the cause of failure in cancer control after radiotherapy. It is known that many types of cancer, that originally have anoxic foci, improve their oxygenation during, and as a result of, fractionated radiotherapy.

Many experimental clinical approaches have been advocated to reduce the radio-resistance of anoxic cells in tumours. Churchill-Davidson, Sanger and Thomlinson first introduced in 1955 the use of hyperbaric oxygen inhalation to improve tumour oxygenation before irradiation with X-rays and this method continues to be evaluated. The use of high LET radiation, such as fast neutrons whose biological effectiveness is much less influenced by oxygen than that of X-rays had been proposed by L. H. Gray and colleagues in 1940. Clinical trials with fast neutron beams are at present in progress in a few centres and soon other accelerated nuclear particles such as negative π-mesons and light atomic nuclei will be available for clinical application.

The scheme of fractionation is often individual to different schools of radiotherapy and there are no generally agreed optimum regimes for different types of cancers. However, the pattern followed to-day is that established by Coutard in 1934 when he developed a 'protracted-fractional' method and there has been little evidence produced since then to suggest that daily fractionation may be disadvantageous. Paterson (1952) was an outstanding pioneer in attempting to establish optimum fractionation, but there is still much need for further detailed study of fractionation schemes in clinical use. Uncertainty about the optimal dose fractionation regime makes assessment of new techniques extremely difficult and the economic consequences of unnecessarily prolonged fractionation are obvious.

It will now be understood that clinical radiotherapy has been practised as a defined medical specialty only since the mid-1920's, when the first deep X-ray therapy machines became available. Radiobiology has been recognized as a distinct scientific discipline for a much shorter time. Since 1940 impressive technological advances in the related fields of physics, chemistry and biology have provided the means for radiobiological investigations on a scale and in detail unattainable before. In this period remarkable progress has been made in our understanding of the action of radiation on cells and tissues. (Lea, 1955; Zimmer, 1961; Elkind and Whitmore, 1967). This book attempts briefly to describe these discoveries and to relate their implications for clinical radiotherapy.

REFERENCES

Becquerel, H. (1896) On various properties of the uranium rays. *Comptes rendus hebdomadaires des séances de l'Académie des Sciences,* **123,** 855-861.

Bergonie, J. & Tribondeau, L (1906) Action des rayons X sur le testicle. *Archives d'électricité médicale,* **14,** 779-791 *et sig.* 911-927.

Coutard, H. (1934) Principles of X-ray therapy of malignant disease. *Lancet,* ii, 1-8.

Desjardins, A. U. (1931) Action of roentgen rays and radium on normal tissues. *American Journal of Roentgenology,* **26,** 1-90.

Elkind, M. M. & Sutton, H. (1959) Radiation response of mammalian cells grown in culture I. Repair of X-ray damage in surviving Chinese hamster cells. *Radiation Research,* **13,** 556-593.

Elkind, M. M. & Whitmore, G. F. (1967) *The Radiobiology of Cultured Mammalian Cells.* New York: Gordon & Breach.

Gray, L. H. Mottram, J. C.; Read, J. & Spear, F. G. (1940) Some experiments upon biological effects of fast neutrons. *British Journal of Radiology,* **13,** 371-375.

Gray, L. H.; Conger, A. D.; Ebert, M.; Hornsey, S.; Scott, O. C. A. (1953) The concentrations of oxygen in dissolved tissues at the time of irradiation as a factor in radiotherapy. *British Journal of Radiology,* **26,** 638-648.

Hodges, P. C. (1964) *The Life and Times of Ernie Grubbe.* Chicago: University of Chicago Press.

Howard, A. & Pelc, S. R. (1953) Synthesis of deoxyribonucleic acid in normal and irradiated cells and its relationship to chromosome breakage. *Heredity.* Supp. No. **6.** 261-273.

Lea, D. E. (1955) *Actions of Radiations on Living Cells.* Cambridge: University Press, 2nd edn.

Muller, H. J. (1927) Artificial transmutation of the gene. *Science.* **66,** 84-87.

Paterson, R. (1952) Studies of optimum dosage. *British Journal of Radiology,* **25,** 505-516.

Puck, T. T. & Marcus, P. I. (1956) Action of X-rays on mammalian cells. *Journal of Experimental Medicine,* **103,** 653-666.

Regaud, C.; Coutard, H.; & Hautant, A. (1922) Contribution en traitement des cancers endo larynges par les rayons. *Xth International Congress. d/Otology,* 19-22.

Roentgen, W. C. (1895) Uber eine neue Art von Strahlen. Erste Mitteilung. *Sitzungsberichte der Physikalische-Medizinischen Gesellschaft in Würzburg,* 132-141.

Warren, W. S. (1928) The physiological effects of radiation upon normal body tissue. *Physiological Review,* **8,** 92-129.

Zimmer, K. G. (1961) *Studies in Quantitative Radiation Biology.* (Translation. H. D. Griffith) Edinburgh; Oliver & Boyd.

2. Biophysical Events

In this book we assume a knowledge of basic radiation physics and that the reader has made a study of the principles involved in the generation of ionising radiations (e.g. in a textbook like *Fundamental Physics of Radiology* by Meredith and Massey, 1972). Radiation energy may be dissipated in the process called *excitation* in which an electron is raised to a higher energy level. This is all that happens with the longer wavelength radiation in the ultra-violet range (Fig.2.1). When the radiation has an even shorter wavelength, its energy is transferred by both excitation and *ionisation,* ionisation being the removal of an electron from its atom or molecule. This chapter will be primarily concerned with those biophysical events which are the consequences of the interaction of ionising radiations and matter.

The radiations and their interactions with matter

X-and γ-rays

Still the most commonly used in clinical radiotherapy, these are electromagnetic radiations consisting of streams of energetic photons (or packets of energy) which can cause ionisation. X-rays are produced when a stream of fast electrons is stopped in a block of metal, usually of tungsten. The energy of the resultant electro-magnetic radiation depends upon the energy of the electrons, as well as the atomic number of the metal. Following interaction of these two factors, photons with a range of energies are produced. The resultant beam of X-rays will therefore have a range of wavelengths but each interaction results in radiation with a 'characteristic wavelength'.

Figure 2.1 shows a comparison of the wavelengths of various sources of radiation on the left, with the diameters of various biological objects on the right. X- and γ-rays fall at the lower end of this diagram, having very short wavelengths. The 'characteristic wavelength' determines the peak energy of the radiation beam generated by a particular X-ray tube and this energy is described in the units of peak kilovoltage, kVp. Because γ-rays result from discrete nuclear disintegrations they have a single energy. In all other respects, X- and γ-rays have similar properties when they interact with atomic matter. From a biological viewpoint the tissues of the body include 'targets' of varying atomic number. Furthermore the mechanisms of interaction between radiation and these 'targets' will apply to ionising radiations of all types, including neutrons and accelerated charged particles which deposit their energy by a nuclear interaction which also depends upon the structure of the target atoms.

The transfer of energy from electro-magnetic radiation to matter is effected by one or more of three processes of attenuation which depend upon the energy

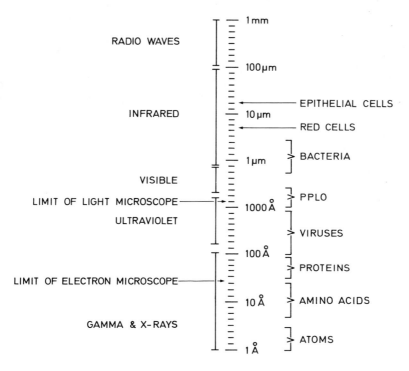

Fig. 2.1 Wavelength — Sources of radiation and targets.

spectrum of the radiation. At energies less than 0.5 MeV the predominant method of interaction is by the *photoelectric effect* in which the photon is completely absorbed by the target atom; an electron is emitted and 'characteristic' radiation is produced. This process is of less interest and importance to radiotherapists now that megavoltage X-ray generators and γ-ray sources (with a peak energy greater than 1 MV) have largely replaced orthovoltage equipment in clinical practice. Compton scattering and pair production are then the main types of interaction. *Compton scattering* is predominant over the energy range 0.5 MeV to ~ 5 MeV and is illustrated in Figure 2.2 which shows the neutrons and protons which form the nucleus of an atom together with the planetary electrons. In the Compton process the incident photon collides with a planetary electron, a recoil electron is produced and the 'scattered' photon leaves with diminished energy. Depending upon its energy this photon may then interact with additional target atoms by further Compton scattering or by the photoelectric effect.

Pair production begins to occur at photon energies in excess of 1.02 MeV but quantitatively it only begins to be of biophysical importance with megavoltage X- and γ-irradiations above 20 MeV peak energy. The incident megavoltage photon is converted into an electron and a positron. The latter is a positively charged electron which is eventually annihilated by collision with an ordinary negative electron to produce two photons of energy 0.51 MeV which are called annihilation radiation.

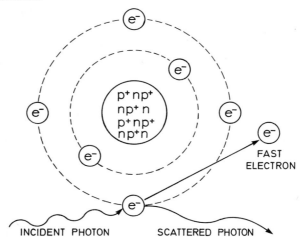

Fig. 2.2 Compton scattering (from Hall, 1973).

Electrons

External electron sources are also used in clinical radiotherapy, and the attenuation of electrons is clearly an important process because the three methods of absorption of photons just described, all involve the production of electrons. These particles (called beta-particles in the context of radioactive disintegrations) are negatively charged and have very small mass. Because of this, electrons will be easily deflected from their track by other electrons. A tortuous track will result so that the *range* of an electron (or the depth to which it penetrates) will be much less than its true track length. What is important to note, however, is that the greatest density of ionisation occurs at the end of this track. The velocity of the electron falls as its energy is degraded to a few tens of electron volts and then the specific ionisation (measured in ion pairs/cm air) rises accordingly. This is because of the increasing probability of interaction between the atoms of the target material and the electrons when they are travelling slowly. In the case of high energy electron beams from a megavoltage betatron or a linear accelerator, the majority of the ionisation will occur at an appreciable depth in the volume of tissue being irradiated; whereas the 0.53 MeV beta particles from a ^{90}Strontium source will deposit most of their energy within a depth of one or two millimetres.

Alpha particles

These are positively charged since they consist of two protons and two neutrons (i.e. they are helium nuclei without the two electrons associated). Because alpha particles are relatively massive (8000 times heavier than an electron) they move more slowly through tissue and penetrate only a short distance; a

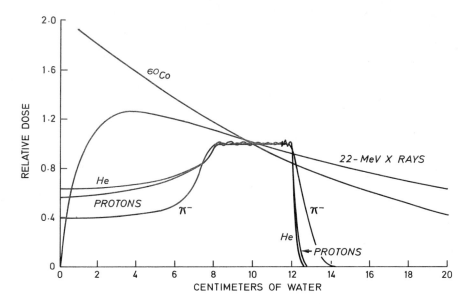

Fig. 2.3 Central axis depth-dose distributions of ^{60}Cobalt γ-rays, 22 MeV X-rays, protons, helium ions and negative π-mesons normalised to 50% dose at 10 cm. (from Raju and Richman, 1972).

few hundred microns at most. By itself, alpha radiation is of little importance in radiotherapy but since the mixed emission from many radionuclides includes alpha particles their mode of action merits description. Furthermore, they are produced by fast neutrons and negative π-mesons in tissue. As with electrons, the greatest density of ionisation occurs at the end of the alpha ray track but this will be short and straight (in contrast to an electron track). This is because the low velocity and the double charge permits more ionisation to occur. Thus an alpha particle is very densely ionising. In terms of *Linear Energy Transfer* units of keV per micron length of track, the average value along the track would be about 100, as compared to 3.0 for orthovoltage X-rays and 0.3 for cobalt γ-rays.

Such values for LET should only be used as a rough guide for purposes of comparison. They may be described as 'track average' LET values (a terminology which has been found especially useful for comparing fast neutron beams) but the ionisation density for alpha particles varies considerably along their tracks, short though they are. This is shown in Figure 2.3 which also illustrates the same phenomenon for other types of charged particle as well. The level of ionisation rises slowly at first, but then there is a final sharp rise to a peak followed by an equally sharp fall to zero when all the energy has been dissipated. At that point all alpha particles will have attracted two electrons to themselves and come to rest in the tissue as helium atoms. This phenomenon has little direct relevance to radiotherapy because alpha particles have such poor powers of penetration in tissues but the specific ionisation curves in Figure 2.3 illustrate a principle which will prove important when the clinical applications of negative π-mesons and other accelerated charged particles are described in Chapter 15.

Neutrons

These particles have no electric charge and will not disturb positively or negatively charged material as they pass through it. No ionisation is produced directly. For this reason interaction can only result from direct collisions with atomic nuclei. In radiotherapy we are concerned with beams of fast neutrons which have sufficient energy to achieve a useful depth dose distribution. While slow or thermal neutrons (with energies less than 100 eV) enter atomic nuclei and are 'captured', fast neutrons (with energies greater than 20 keV) interact mainly by elastic collisions with the nuclei. The most efficient interaction will occur when there is a head on collision with a proton (i.e. a hydrogen nucleus) because protons and neutrons have equal mass. All the neutron energy will therefore be transferred to the proton which will recoil (or be 'knocked on').

Most tissues have a high density of hydrogen atoms so that interaction with a fast neutron beam results in the local production of recoil protons with up to the same initial energy. Thus, heavy positively charged particles are produced in the irradiated volume. These will be quickly slowed down and then become very densely ionising in a similar way to that described for alpha particles and with a high LET, up to 90 keV/μm at the end of the proton track, and averaging 10 to 50 keV/μm depending on the energy of the neutrons.

Elastic scattering also occurs with nuclei of oxygen, carbon and nitrogen from which densely ionising recoil particles are produced. Inelastic scattering of fast neutrons may result from interaction with heavier atomic nuclei in tissues with the production of γ-rays. Two other forms of attenuation may take place; nuclear disintegrations and neutron capture. In the process of nuclear disintegrations neutrons are absorbed into nuclei resulting in such instability that they explode releasing alpha particles, deuterons, protons and other neutrons. Neutron capture may take place in hydrogen and in nitrogen atoms making them radioactive and emitting γ-rays and protons, respectively.

Heavy charged particles

We have seen that fast neutrons may produce recoil protons by elastic collisions with hydrogen nuclei, and that collisions between neutrons and carbon, nitrogen, oxygen and other nuclei in a tissue, will produce charged particles which will also be densely ionising. Alternatively, heavy charged particles may be accelerated by a very high energy machine (e.g. a synchrotron) to have enough energy to penetrate a tissue to a useful depth for therapeutic purposes before ionisation reaches a peak. This is shown in Figure 2.3 where protons, helium ions, and negative π-mesons are seen to deliver a relatively higher dose at 10 cms depth than at the 'surface' in contrast to the steady fall in depth dose with orthodox radiation sources. Of additional importance is the fact that the dose delivered at 'the depth' by these heavy charged particles is biologically more effective because they are more densely ionising than orthodox radiation. This will be discussed again in Chapter 15, Figure 15.5.

LET and RBE

Linear Energy Transfer (LET) has already been mentioned in this chapter in the context of those modes of radiation which are more densely ionising than X- or γ-rays. It is obviously difficult to give a simple description of LET because of the very nature of the phenomena described in those earlier sections; where a neutron or a charged particle was followed along its track through a tissue. The energy is decreasing and the probability of interaction between particle and target atom is increasing with greater distance along the track. Even along as short a length of track as one micron (μm) the density of ionisation will vary — not to the extreme extent of that shown in Figure 2.3, but to some extent. A simplified parameter is therefore chosen, namely the track average LET expressed in units of energy released per length of track (keV/μm). This may still be an over-simplification of the phenomenon but the parameter will be used for comparative purposes in this book.

Figure 2.4 is a classical diagram which shows a biological target with examples of the linear distribution of ionising events across its diameter. More densely ionising radiation is seen to have a greater probability of hitting the target one or more times as it traverses that distance. Less densely ionising radiation is less likely to do this and the target may not be ionised at all or only once. If the 'target' is as complex as a mammalian cell then some critical part of the cell, such as the nucleus, may need to be hit more than once before any damage will be achieved in the biological sense. The choice then lies between a requirement either for one sensitive site to be hit twice or more, or for one hit to occur in two or more sensitive sites within the cell. This choice of *target theory* is important in fundamental radiobiology.

For clinical purposes it can be assumed that analysis of data from most cell survival curves (see Ch. 6) support the second alternative, namely the *multi-target single-hit* theory. If only one sensitive site is hit then the cell may recover from such physical damage and no biological effect will be achieved. (Recovery from sub-lethal damage is discussed in Chapter 7.) If more than one sensitive site is hit, however, then biological damage will be irreversible. More densely ionising radiation is more likely to produce this effect, as was shown in Figure 2.4. Thus high LET fast neutrons will be relatively more effective in

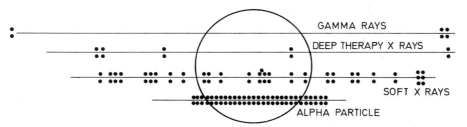

Fig. 2.4 Separation of ion clusters in relation to the size of a biological target (from Gray, 1946).

producing biological damage than low LET γ-rays. The ratio of doses which are needed to produce a given amount of biological damage indicates the Relative Biological Effectiveness (RBE).

For reference purposes it is convenient to compare RBE values with that of a standard source of therapeutic radiation with the lowest RBE. Megavoltage X-rays and cobalt γ-rays then have an RBE value of unity; medium voltage X-rays then have a value of 1.15 but fast neutrons and accelerated charged particles tend to have much higher RBE values of two or more. In the latter case, however, the value for RBE varies with the size of the individual dose or, more particularly, with the level of biological damage used to measure the RBE ratio. This variation in RBE value will be discussed in greater detail in Chapter 7 since it is attributable to the relatively smaller amount of recovery from the sub-lethal damage following high LET radiation. The subject is also considered in Chapter 15, where RBE values will be listed for various tissues exposed to new modalities of therapeutic radiations like fast neutrons and negative π-mesons.

Ionisation and radical formation

There is an apparent paradox in the term 'ionising radiation' which might be expected to mean a particular form of radiation which produces ionisation in target molecules. But many molecules in aqueous solutions already exist, at least partly, in an ionised state due to simple dissociation into positively and negatively charged ions which coexist in a stable equilibrium. Thus Na Cl dissociates into Na^+ and Cl^- ions just as water itself dissociates into H^+ and OH^- ions. The consequence of X-irradiation, however, is the formation of pairs of abnormal ions (called free radical ions) which are *not* in equilibrium. Thus, water is ionised by X-irradiation into H_2O^+ ions and free electrons. Such free radical ions are extremely unstable and can form neutral free radicals. *Free radicals* are uncharged atoms or molecules with an unpaired electron in the outer orbit. For example:

$$H_2O^+ \longrightarrow H^+ + OH^{\cdot}$$

(where the dot signifies an unpaired electron). Although the free radicals are more stable they are still very reactive. Both free radical ions and the resultant free radicals disrupt normal molecular structures and damage the biological target. This radiobiological damage is not the consequence of ionisation due to simple dissociation, therefore. The important result of 'ionising' radiation is the production of free radicals. Their formation will now be described in more detail.

When ionisation occurs in air an average of 34 electron volts is dissipated; probably less in liquids and solids. This does not mean that it takes that amount of energy to eject an electron from an atom, however, since that may only need about 10 eV. Two thirds of the energy is dissipated in excitation, which is relatively unimportant in the production of radiation damage in human tissues.

When an electron is knocked out of one of the outer shells of an atom the atom becomes positively charged and is called a positive ion; the electron may then interact with another atom to form a negative ion and the two ions so produced are called an ion pair. Such an ion pair will have a very short lifetime of the order of 10^{-10} sec before producing neutral free radicals which are relatively more stable with a lifetime of up to 10^{-3} sec. It is the free radicals formed from water which are responsible for about half of the biological effects of radiation, so they require more detailed consideration. The other half is due to direct ionisation of critical biological molecules.

Radiation chemists use the technique of *pulse radiolysis* to study the nature and kinetics of radical formation. Typically, an aqueous sample is given a pulse of ionising radiation from a linear accelerator and simultaneously the absorption spectrum of ultra-violet or visible light is measured using a spectrophotometer connected to a cathode ray oscilloscope. Because of the very short life-time of the free radicals, an elaborate timing mechanism is used to synchronise photography of the CRO tracing with the microsecond (or shorter) pulse of electrons delivered to the sample. Analysis of the tracing provides information of the nature of the free radicals produced and their life-time in the aqueous medium. In solid samples (and frozen aqueous samples) the free radicals have a very much longer life-time and they may be studied further using additional techniques such as *electron spin resonance.* In this technique the sample is placed between the poles of a powerful magnet and determination of the perturbation of the magnetic field indicates the nature of the reactive chemical species produced by the ionising radiation. All this involves an important field of scientific study which is beyond the scope of this book (a detailed review is given in the textbook *Radiation Chemistry* by A. J. Swallow, 1973).

Following the primary absorption of radiation energy a complicated series of events will lead to the breakage of a number of chemical bonds. Since biological material consists of 70 to 90 per cent water, the radiation chemistry of water is obviously of great importance. The primary event is the ionisation of the water molecule to give a positive ion and a free electron. This is followed by a complex series of reactions leading to chemical products some of which are listed in table 2.1 with their G values. The symbols $OH^{\cdot}$ and $H^{\cdot}$ denote free radicals as distinct from the ions HO^- and H^+. The $OH^{\cdot}$ and $H^{\cdot}$ radicals are formed by the breaking of one of the H-O bonds in the water molecule and are extremely reactive. e^-aq stands for the so-called 'hydrated' electron which is a free electron trapped inside a 'cage' of water molecules. Hydrated electrons are surpris-

Table 2.1   Yields of products in irradiated water at neutral pH

Product	Yield, G
$OH^{\cdot}$	2.6
e^-_{aq}	2.6
$H^{\cdot}$	0.6
H_2O_2	0.75
H_2	0.45

(G is the number of molecules produced by the absorption of 100 eV energy.)

ingly stable with a life-time in pure water of several microseconds. They may decay by combining with hydrogen ions (H^+) to form more hydrogen radicals ($H^{\cdot}$) but in biological systems are most likely to react chemically with dissolved oxygen or organic molecules. Many other radicals are formed but it is only proposed to discuss radiobiologically important types of reaction such as those in which aqueous free radicals react with organic biological molecules and with oxygen. From a biological point of view it makes little difference whether a molecule is damaged directly or indirectly although the aqueous nature of most biological targets in man makes the indirect mechanism equally as important as direct action.

If the symbol RH is used to represent an organic molecule in human tissue then the *direct effect* of ionising radiation upon such a molecule may be represented by the following equation:

$$RH \xrightarrow{\text{radiation}} RH^+$$
$$RH^+ \longrightarrow R^{\cdot} + H^+$$

Direct interaction has ionised the molecule leading eventually to the production of the free radical $R^{\cdot}$.

The *indirect effect* will also operate. This involves the production of the aqueous free radicals $OH^{\cdot}$ and $H^{\cdot}$ already described above as following the ionisation of the water content of most human tissues. Typical equations which may now apply to the interation of these radicals with organic molecules in their immediate vicinity are:

$$RH + OH^{\cdot} \longrightarrow R^{\cdot} + H_2O$$
$$RH + H^{\cdot} \longrightarrow R^{\cdot} + H_2$$

The reactions of hydrated electrons will also give rise to organic free radicals. $R^{\cdot}$ radicals may give rise to permanent damage, but the damage could be repaired, say, by reaction with an SH compound:

$$R^{\cdot} + -SH \longrightarrow RH + -S^{\cdot}$$

($-$ where the $-S^{\cdot}$ radical is inert.)

Figure 2.4 illustrates how the relative importance of these direct and indirect effects will vary with the LET of the radiation, i.e. the distribution of ionising events across the target and the probability of these ionisations occurring inside or outside a critical volume. It can be seen that a biological target is more likely to be damaged indirectly by radiation of low LET. Furthermore, the direct effect is not only more likely to operate with high LET radiation, but this will also be relatively much more damaging, dose for dose.

The oxygen effect.

In table 2.1 the chemical products of the ionisation of water which had the highest yield were the hydroxyl radical $OH^{\cdot}$ and e^-aq, the hydrated electron. The hydroxyl radical has one oxidising equivalent whereas the hydrated electron has one reducing equivalent and when molecules are present in water which are oxidisable or reducible they may be attacked by the corresponding radicals. If oxygen is present in the irradiated tissue then an increased amount of damage might be expected to be produced through the following reaction:

$$R^{\cdot} + O_2 \longrightarrow RO_2^{\cdot}$$

This is an organic peroxy radical which cannot easily be repaired so that the radiation damage is fixed. In certain parts of cells a chain reaction may even be generated which would involve more of the original biological molecule RH:

$$RO_2^{\cdot} + RH \longrightarrow RO_2H + R^{\cdot} \text{ (and so on.)}$$

Another damaging process may involve the formation of *hydrogen peroxide* in the irradiated tissue if oxygen reacts with the aqueous free radical $H^{\cdot}$:

$$H^{\cdot} + O_2 \longrightarrow HO_2^{\cdot}$$
$$2HO_2^{\cdot} \longrightarrow H_2O_2 + O_2$$

H_2O_2 may also be formed in a similar manner from hydrated electrons. Hydrogen peroxide is very toxic to biological structures and is a relatively stable molecule.

In summary, the oxygen effect in radiobiology involves a competition between dissolved oxygen and endogenous hydrogen donors in the particular tissue. This competition will determine the amount of radiation damage and obviously depends upon the concentration of oxygen molecules in the vicinity of the target molecule. It will be affected by any change in the balance of reducing or oxidising (e.g. H_2O_2) species in the immediate environment which will tend to protect or sensitize the tissue to irradiation of low LET. The oxygen effect is reduced as the LET increases, because with high LET radiations the damage is much more severe than at low LET, and cannot be increased still further.

Protection and sensitization

By far the best clinical radiation sensitizer is oxygen dissolved in the tissues at a physiological concentration. Under normal conditions, of course, oxygen is usually present in the cellular environment so that sensitization can only be demonstrated by comparison with the situation when tissues are (presumably abnormally) hypoxic. If the concentration of oxygen molecules in the vicinity of the irradiated tissue is reduced, this will swing the radiation chemical competition in favour of the endogenous hydrogen donors in that tissue.

The balance may be affected in the opposite direction by the addition of protective compounds such as those containing the sulphydryl group $-SH$ (e.g. cysteamine and cysteine). In terms of reactions, then, while oxygen is more damaging because of the effect already discussed:

$$R^{\cdot} + O_2 \longrightarrow RO^{\cdot}_2$$

a compound containing an $-SH$ group will have the following effect:

$$R^{\cdot} + -SH \longrightarrow RH + -S^{\cdot}$$

The sulphydryl group has enabled the ionised molecule to be restored to its normal state. In practice, most $-SH$ compounds investigated so far have proved toxic in the concentrations necessary to be effective in clinical use. A new class — the thiophosphates — are not toxic however, but are not protective until they are transformed by enzymes inside living cells. Most cells contain measurable amounts of $-SH$ groups, but the concentrations are not high enough to compete successfully with atmospheric O_2 for the $R^{\cdot}$. As well as protection by the mechanisms discussed there are also several other ways by which protective compounds may operate:

Alternative sensitizers to oxygen must not only act by intracellular binding of naturally occurring radioprotectors (like the $-SH$ compounds) but may also act by swinging the radiation chemical competition in favour of the oxidative pathway and against the reductive. The main products of the irradiation of water (Table 2.1) are the $OH^{\cdot}$ radicals which are oxidative, and the hydrated electrons (e^-aq) which are powerful reducing agents. Any compound that is *electron-affinic* will tend to be radiosensitising. This is a class of compounds (e.g. metronidazole and other nitro-imidazoles) which may have clinical applications (see Ch. 15) since they have low toxicity and can be expected to mimic the oxygen effect in those hypoxic tissues to which they can diffuse. Compounds which interfere with DNA synthesis may also be sensitisers (see Ch. 4) and so may cytotoxic chemotherapeutic agents used in conjunction with radiotherapy. Strictly speaking, such compounds should be used at a non-toxic level if true radiosensitization is to be shown (cf. oxygen). Most therapeutic regimes of combination chemotherapy and radiotherapy show only an additive effect from the two cytotoxic modalities.

In summary, the biophysical effects of those forms of radiation in clinical use involve the attenuation of either an incident X-ray photon (producing a fast electron) or an incident fast neutron (producing a recoil proton) in tissues leading to the production of ion pairs and then free radicals. These interactions lead to the breakage of chemical bonds and biological effects which are the subject of the next chapter.

REFERENCES
Gray, L. H. (1946) Comparative studies of the biological effects of X-rays, neutrons and other ionizing radiations *British Medical Bulletin*, **4,** 11-18.
Hall, E. J. (1973) *Radiobiology for the Radiologist* London: Harper and Row, p.9.
Raju, M. R. & Richman, C. (1972) Negative pion radiotherapy: physical and radiobiological aspects. *Current Topics in Radiation Research*, **8,** 159-233.

FURTHER READING
Meredith, W. J. & Massey, J. B. (1972) *Fundamental Physics of Radiology,* Bristol: John Wright.
Swallow, A. J. (1973) *Radiation Chemistry: An Introduction* London: Longmans.

3. The Biological Target

The tissues of the adult human body contain about 5×10^{13} cells. These cells are the basic biological units and they represent the biological target of ionising radiation. In the previous chapter the biophysical events leading up to the absorption of radiation damage and the immediate effects of such ionisation were described. The interaction of radiation with the biochemical constituents of the body will be considered in the next chapter. In this chapter the different types of human cell will be discussed in relation to their response to radiotherapy.

All forms of cancer therapy would be simplified if some specific difference were to be found between malignant and normal cells. An antibiotic might then be developed which would be lethal to the abnormal cells with minimal effect on the normal tissues. The therapeutic ratio would be large. For the present, the difference between the two types of cells lies more in their behaviour in a tissue than in anything intrinsic which can be recognised in individual cells. (The lack of any systematic difference in radiosensitivity at the cellular level will be discussed in Chapter 6.)

Properties of malignant cells

Those few differences which have been recognised have not yet proved useful in radiotherapy. The surface properties of some malignant cells differ from the normal and this may explain the spread of tumours. The nuclei of tumour cells in carcinoma of cervix, for example, have a greater volume than those of the normal cells in that tissue. At the sub-cellular level such tumour cells have fewer ribosomes per unit area of cytoplasm and show a reduction in the amount of intercellular space. None of these differences has any direct application to radiotherapy, however.

The discovery of systematic intrinsic differences between malignant and normal cells is more likely to lead to advances in chemotherapy and immunotherapy than in radiotherapy. This is because the random distribution of ionising events throughout the cells in an irradiated tissue removes the possibility of delivering damage to any discrete targets which may be identified in malignant cells. For the present it is the behaviour, particularly the growth pattern, of the cells which provides a guide to the most efficient use of radiotherapy in its prime application, namely the treatment of circumscribed volumes of tissue.

The Law of Bergonié and Tribondeau

When Bergonié and Tribondeau were looking into the effect of radiation on the rat testes they discovered (1906) that the dividing (germinal) cells were markedly affected by the radiation, whilst the non-dividing (interstitial) cells appeared undamaged. On the basis of these observations they derived their Law which states that cells are radiosensitive if they are mitotically active, if they normally undergo many divisions and if they are morphologically and functionally undifferentiated. Thus the radiosensitivity of a tissue is directly proportional to its mitotic activity and inversely proportional to the degree of differentiation of its cells.

This Law has led to the generalisation that actively dividing tissues are 'radiosensitive' and non-dividing tissues are 'radioresistant'. Thus in mammals, the liver is often classified under 'radioresistant tissues' since in the adult it exhibits little active cell division and is composed of specialised cells. In contrast, the cells of the epithelium of the intestine are classified under 'radiosensitive tissues' (Table 3.1). Such generalisations are misleading since it is essentially the processes of cell division that are 'radiosensitive' and not the different

Table 3.1   Cell populations and their kinetic properties (After Bertalanffy and Lau, 1962)

No mitosis No cell renewal	Low mitotic index Little or no cell renewal	Frequent mitoses Cell renewal
C.N.S. Sense Organs Adrenal Medulla	Liver Thyroid Vascular endothelium Connective tissue	Epidermis Intestinal Epithelium Bone Marrow Gonads

types of cells in tissues. The only cell types that are exceptions to this rule are the oocyte and the small lymphocyte which do not divide but which are nevertheless radiosensitive.

For a valid comparison of radiosensitivity to be made between two types of cell, the same criteria must be used for both. The Law of Bergonié and Tribondeau would conclude that the liver cells, which rarely divide, are radioresistant, so that if, after 1000 rads, the state of the liver was compared with that of the intestinal epithelium, the former would appear intact, the latter would be damaged and the law would apparently be confirmed. This is because the rapidly dividing epithelial cells of the intestinal crypts suffered damage to their reproductive ability leading to defects in the structure of the intestinal wall; while the static, occasionally dividing, liver cells remained morphologically intact. On the other hand, if the liver cells are stimulated to divide by partial hepatectomy the same criterion of damage can be compared in both the liver cell and the intestinal epithelial cell, and the doses needed to give the same effect are found to be very similar.

Radioresponsiveness and radiosensitivity

Thus, it is the processes of cell division that are 'radiosensitive' rather than the functional integrity of the cells in an irradiated tissue. The alternative word would be *radioresponsive,* so long as the word is used to describe the eventual fate of the irradiated cells. Some tissues, normal and malignant, take longer to respond to radiotherapy than others. The immediate response will appear slight but it is the end result that concerns the radiotherapist. Tissues that respond quickly are often described as radiosensitive, the others as radioresistant. In terms of ultimate radioresponsiveness, however, there may be little or no difference. Figure 3.1 illustrates this for a number of normal mammalian cell populations which, because they vary in the transit time of the mature cells, show a variation in the period of time before they respond to radiation. Just because the intestinal epithelium responds most quickly, is no reason to state that it is the most 'radiosensitive'. The Law of Bergonié and Tribondeau has

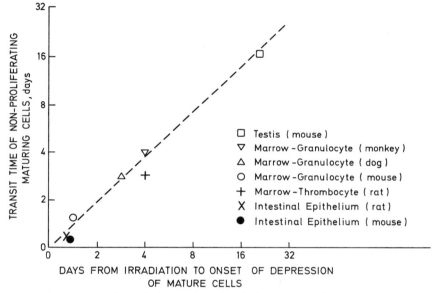

Fig. 3.1 Relation of transit time to time of radiation response. (from Patt, 1963).

sometimes been stated: the cell systems that have a high rate of cell division generally suffer the most radiation damage, while the tissues that show little active cell proliferation suffer proportionately less. But that statement is misleading because the clinician needs to consider the *ultimate* response of the irradiated tissue and, unless a new definition such as radioresponsive is to be used, then the old definition 'radiosensitive' needs to be understood in terms of the probability that mitotic death will occur when the cells next enter division; whenever that may be.

Any cell of the body has only a statistical probability of being damaged by radiation and some cells will remain unscathed. Whether or not the damage becomes manifest as cell death depends to a very large extent on the detailed characteristics of the renewal system of which that cell is a part. Cells of rapidly renewing systems will manifest this damage as cell lethality within hours or weeks, at dose levels used in clinical practice. Cells of non-renewal systems may never manifest the 'latent' damage unless artificially forced to act as a renewal system (e.g. regenerating liver). Table 3.1 showed examples of how the cell populations of the parenchymal portions of various tissues and organs can be classified with respect to their kinetic properties. It will be seen that such a classification has an important bearing on the radiation response of such tissues and organs.

All tissues are composed of parenchymal cells, which carry out the function of the tissue, plus supportive elements (e.g. connective tissue) and the blood vessels which transport metabolites to the parenchymal cells. If radiosensitivity is defined in terms of cell disappearance or death then postmitotic cells (e.g. those of the C.N.S.) must be defined as radioresistant, while most of the vascular and connective tissue cells are intermediate in sensitivity.

The radiosensitive tissues (or those that respond most quickly) are those in the right-hand column of Table 3.1 (e.g. the epidermis, mucosa of the gastrointestinal tract, or the blood-forming tissues). With such tissues the early radiation effects, particularly after low radiation doses, depend primarily on destruction or mitotic inhibition of parenchymal cells. The disappearance of parenchymal cells may be rapid but the cells are usually replaced within a short time. Accordingly, there may be early damage, but quick recovery.

Vascular tissue

If the exposure is high enough to produce vascular changes leading to narrowing or occlusion of small blood vessels, there may be a delay in the replacement of parenchymal cells which are dependent on an intact blood supply. In addition, changes initiated in the vascular or connective tissue (vasculoconnective tissue) are often not repaired in a typical fashion, and this damage may actually increase in severity in the months and years following radiation. Therefore, long after irradiation there may be secondary changes in parenchymal cells which are a result primarily of the vascular damage. Such long-term changes will be considered in Chapter 11.

Furthermore, in tissues which contain parenchymal cells with a low renewal rate such as liver, kidney, muscle, or brain, high doses have no immediate apparent effect on the parenchymal cells, but only moderate doses are needed to damage the vasculature. Thus, the major changes seen in the parenchymal cells following moderate doses of irradiation are late effects and are secondary to changes in the vascular tissue. Massive radiation doses are required to produce an acute lethal effect in the parenchymal cells of such tissues. It is difficult to assess the indirect contribution of changes in the vascular and connective tissue following massive radiation exposures.

The radiation response of an organ can also be markedly influenced by the condition of other irradiated or nonirradiated tissues and cells in the body due to so-called *abscopal* effects. If certain parenchymal cells are removed but can be replaced by differentiation of more primitive cells or by migration of similar cells from a different tissue, the overall effect may be lessened. For example, bone marrow cells from one part of the body can repopulate irradiated marrow in other parts of the body.

Damage to the vascular system contributes indirectly to many of the changes in other tissues. Some aspects of this interaction will be described in Chapter 5 in connection with the growth rate of tumours. In functional terms the heart and the large arteries and veins appear radioresistant while the capillaries seem to be radiosensitive. This is because occlusion occurs in capillaries and small arteries after moderate doses of irradiation. Several mechanisms may be involved in this occlusion: the narrow capillary lumen may be blocked if endothelial cells swell or respond to irradiation by hyperplasia. Damage to the endothelial surface may lead to clotting which would tend to block the smaller vessels, and such vessels may also collapse due to external pressure from an inflammatory reaction in the vicinity. Capillary occlusion will block the blood supply not only to the tissues in the immediate vicinity of the occluded area but to all tissue further along the capillary. For this reason capillary endothelium is perhaps the most important tissue limiting radiation to a patient.

Normal tissue

At the cellular level the endothelial lining of large vessels is probably as sensitive as that in capillaries but, because of the large diameter of those vessels, even if endothelial proliferation and swelling or blood clotting does occur, the vessels will not be occluded. The end result of irradiating a human tissue is illustrated in Figure 3.2 which compares (a) normal skin with skin removed from two patients who had received (b) a tolerance dose of radiation five years before and (c) an overdose of radiation twenty years previously. In addition to the radiation damage, which will be described later, these photographs serve to illustrate an important principle of histopathological specimens, namely the fixity of the pictures. Just in order to obtain these microscopic pictures the tissues have had to be fixed, sectioned and stained. The dynamic aspects of a viable tissue and the kinetic properties of its constituent cells must then be lost; except in so far as the specimen is one example in space and time of those

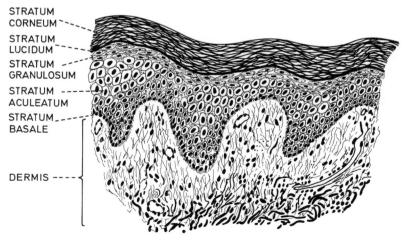

Fig. 3.2(a) The features of normal skin.

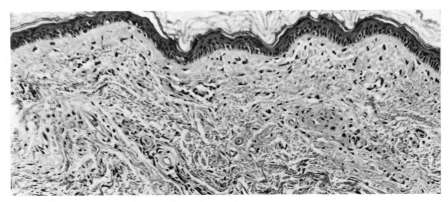

Fig. 3.2(b) Skin after a tolerance dose of radiotherapy.

factors in that tissue. In the remaining chapters of this book the dynamic aspects of radiobiology will be emphasized but there is important information that can be derived from the histopathology of a fixed tissue, especially when consideration is given to the period of time which has elapsed since the tissue was irradiated. The ageing of cells and tissues will be discussed in Chapter 12 in the context of the possible ageing effect of ionising radiation.

The histopathological changes in an irradiated tissue are illustrated by comparison of the three illustrations of skin sections shown in Figure 3.2. The first picture shows normal skin undamaged by radiation, with the multi-layered epidermis and all the dermal structures, including an abundant blood supply. In the middle picture the skin had received a 'tolerance dose' of radiation as treatment for a rodent ulcer five years previously. Some damage is evident as a mild hyperkeratosis with an increase of melanocytes in the basal layer of the epidermis. In the dermis there is swelling of the capillary endothelium and the larger vessels show obliterative thickening. Although there is a sparsity of the more differentiated dermal structures there is an apparent increase in cellularity in the form of fibroblasts with swollen nuclei. The tissue has indeed 'tolerated' the absorbed radiation in the sense that the skin can still function

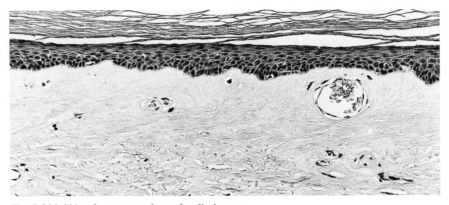

Fig. 3.2(c) Skin after an overdose of radiotherapy.

adequately. On the other hand, damage to the vascular supply has resulted in small changes in architecture and many of the cells that remain have lost their capacity to proliferate, becoming giant in size.

The skin shown in the bottom picture was removed from a woman who received radium treatment for a wart on her hand twenty years previously when she was a child of two. The skin shows severe atrophic changes in the epidermis and almost no cellular structure at all in the dermis except for one large telangiectatic vessel. Other sections showed incipient necrosis in the collagen and a radionecrotic ulcer would be a likely sequel in this tissue because it has lost most of its functional elements.

These photomicrographs illustrated a change in a normal tissue at one point of time after irradiation. If it were possible (e.g. by serial biopsy) to follow the sequence of changes in the histology of such an irradiated tissue then the following would be found to occur: to start with there would be immediate inhibition of division in cells which normally are mitotically active, while chromosome aberrations might be seen in those cells which are in division and many cells would show degenerative changes. Later on, haemorrhage and oedema would be evident, and the most 'radiosensitive' cells (i.e. those which respond earliest) would have degenerated and phagocytosis of cellular debris would have occurred. Following this, the picture would be of regeneration but often with abortive mitosis until the tissue is restored to its final state. This will show the sort of abnormality seen in Figure 3.2(b) with secondary changes in the parenchymal cells and permanent damage to the vascular and connective tissues.

Malignant tissue

The problem in radiotherapy is the recognition of the stage reached in such a post-irradiation sequence. Has the normal tissue stabilised after a tolerance dosage has been delivered or is it still degenerating towards eventual necrosis? Has this tolerance dosage resulted in the successful eradication of malignant tissue or are those malignant cells which are visible in a histopathological specimen viable and not sterilized? No one microscopic section permits an absolute answer to such questions without clinical knowledge of the stage reached in the dynamic sequences following irradiation.

This problem is illustrated in Figure 3.3 which shows photomicrographs of a carcinoma of cervix biopsied before and after radium treatment. Before treatment (Fig. 3.3(a) the biopsy shows a poorly differentiated invasive squamous cell carcinoma in which occasional foci of tumour necrosis and secondary infection are seen. Mitotic figures are not numerous nor is there oesinophilia of the cytoplasm. By contrast, the biopsy taken two weeks after radium insertion (Fig. 3.3(b) shows sheets of tumour cells with a marked alteration in morphology. Giant cells with enlarged nuclei and prominent nucleoli are evident; some of the cells being multi-nucleated. There is increased mitotic activity but much of this is bizarre. Many of the tumour cells show eosinophilia which seems to be one of the hallmarks of radiation damage.

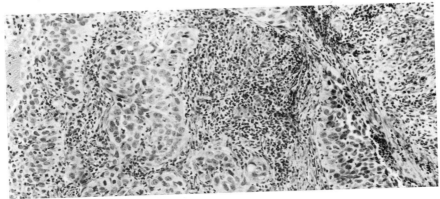

Fig. 3.3(a) Carcinoma of cervix before treatment.

After a period of only two weeks it is just not possible to know whether the histological appearances in Figure 3.3(b) indicate a successful response to radium. If anything, there is an increase in cellularity with large numbers of inflammatory cells visible throughout the tumour masses. While many of the tumour cells already show histological evidence of cell death, there is no way of telling whether or not some of the others may have retained the capacity to proliferate (This question of reproductive death is discussed in Chapter 6). Even the presence of mitotic figures may not necessarily be evidence of active tumour since such mitosis may turn out to be abortive, leading to the phenomenon of 'random death' in the tumour cell population. Some more dynamic evidence is needed, such as failure of the tumour to regress after treatment or regrowth. Thus, histologically intact tumour cells have been observed in animals after a radiation dose which is known to be curative, but such cells fail to reproduce tumours after transplantation (Suit and Gallager, 1964). Details of the cell population kinetics of normal and malignant tissues will be discussed in Chapter 5 but the growth cycle in living cells must be considered now before the biochemical damage of radiation is dealt with in Chapter 4.

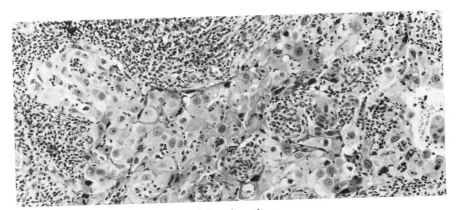

Fig. 3.3(b) Carcinoma of cervix after radium insertions.

Tissue growth

Growth of a tissue will involve a change in the total number of its constituent cells, adding up to a net increase. It must be remembered that the cell population of most normal tissues in a mature animal remains static in total number even though a rapid turnover may continue in some cellular components (e.g. the intestinal epithelium, discussed in Chapter 8). With tumours the net increase in cell number is shown as an increase in volume. The diameter of an accessible tumour must be the commonest of all clinical measurements in cancer work. Increase in tumour diameter is positive evidence of tumour cell proliferation, although the converse can rarely be proved since proliferation may still occur with no apparent change in tumour volume. This will be because a relatively small number of the cells are proliferating. It is these cells, however, which will be seen to influence the response of a tumour (and its limiting normal tissue) to radiotherapy.

When mammalian cells proliferate, the period of time from one mitotic division to the next can be divided into three phases (Fig. 3.4). The phase which is most easily recognized by present techniques is the DNA synthetic or S-phase. DNA synthesis can only be observed during that part of the cell cycle whilst other cellular constituents tend to be synthesised throughout interphase. The gaps before and after the S phase are called respectively the G_1 and G_2 phases. G_1 and G_2 are not only gaps in the cell cycle with respect to DNA synthesis, they are to some extent gaps in our knowledge of what metabolic and other processes occur during those phases. It is known that the synthesis of both

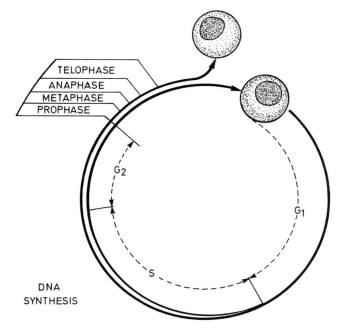

Fig. 3.4 The phases of the cell cycle.

RNA and protein continues throughout interphase but the relationship with DNA synthesis comes more into the fast-moving field of molecular biology. The effect of radiation on the biochemistry of the cell and on synthesis of its main constituents will be considered in the next chapter, however.

REFERENCES

Bergonié, J. & Tribondeau, L. (1906) *See* English translation by Fletcher, G. H. (1959) Interpretation of some results of radiotherapy and an attempt at determining a logical technique of treatment. *Radiation Research,* **11,** 587-588.

Bertalanffy, F. D. & Lau, C. (1962) Cell renewal. *International Review of Cytology,* **13,** 359-366.

Patt, H. M. (1963) Quantitative aspects of radiation effects at the tissue and tumour level. *American Journal of Roentgenology, Radio Therapy and Nuclear Medicine,* **90,** 928-937.

Suit, H. D. & Gallager, H. S. (1964) Intact tumour cells in irradiated tissue. *Archives of Pathology,* **78,** 648-651.

4. Biochemical Damage

To understand the biochemistry of the mammalian cell one has to imagine an industrial concern, or factory, in which a number of jobs are performed under the same roof. During normal working conditions (e.g. before the disruptive effects of ionising radiation) the whole concern is working in harmony with each job progressing at an appropriate rate. Some jobs are very complicated, some are simple. Disruption of certain of the jobs will bring the whole concern to a stop. Others are less important and normal working may still continue without them for a period whilst repairs are undertaken. Figure 4.1 shows a cell which can be considered as the 'industrial concern' with its various subcellular organelles undertaking the different jobs required for the efficient working of the cell.

All mammalian cells have certain common features of internal organisation such as a functional nucleus (the mature erythrocyte is the exception to this rule) which is separated from the cytoplasm by the nuclear membrane except at the time of cell division. Other common features include the mitochondria for the production of energy, ribosomes for the synthesis of proteins, the Golgi apparatus which is active in secretory cells, the endoplasmic reticulum where secretory proteins are synthesised, and the lysosomes which represent the digestive system of the cell. Surrounding all this is a lipoprotein membrane

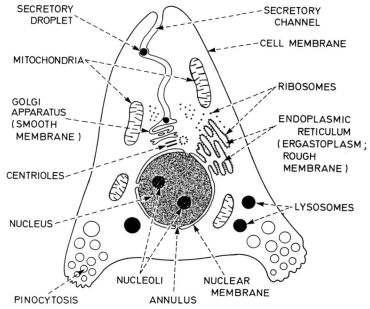

Fig. 4.1 Organelles of a 'typical' cell.

whose integrity is just as important to the smooth functioning of a cell as would be the outer walls to the smooth running of a factory. Damage to the wall or to the other sub-cellular organelles in the cytoplasm will certainly be disruptive but the most sensitive macromolecules or structures are contained within the nuclear membrane: the nucleoli and the nucleus itself.

Radiosensitivity of the nucleus

This has been shown by two sorts of experiment. In one case the alpha irradiation from Polonium was delivered either to the cytoplasm or to the nucleus of cells in tissue culture by the exact positioning of the tip of a fine needle upon which the Polonium had been deposited. A dose in excess of 25 000 rads delivered to the cytoplasm had no effect upon the proliferation of these cells while the *mean lethal dose* to the nucleus was less than 150 rads. This experiment shows that the nuclear region of the cells is at least 100 times more radiosensitive than the cytoplasm. The other experiment compared the radiation effect of tritium (^{3}H) labelled water with that of tritiated thymidine. Thymidine is a specific requirement for DNA synthesis and so tritiated thymidine will be localised in the chromosomes in the cell nucleus, while tritiated water should be evenly distributed throughout the whole cell, both nucleus and cytoplasm. The result of that experiment showed an even greater difference in radiosensitivity, in that more than 1000 times the radioactivity from tritiated water was required to equal the amount of cell damage produced by tritiated thymidine. This sort of experiment fails to ensure a uniform uptake of isotope into the various molecular structures of the cell, however. The sensitivity of its DNA should more properly be compared with that of its RNA and protein molecules in terms of the actual incorporation of radioactive atoms. This is a complex problem which will be considered in the next section on molecular radiobiology, under the heading of nucleic acids.

Molecular radiobiology

Radiation effects at the molecular level can best be understood in the setting of what is now a classical part of the central dogma of molecular biology (Fig. 4.2) which states that RNA is transcribed from DNA. The genetic information of the cell is transferred from the nucleus by *messenger RNA* to ribosomes in the cytoplasm. Protein synthesis then occurs at the site of the ribosomes where the necessary amino-acids are assembled by *transfer RNA* using the

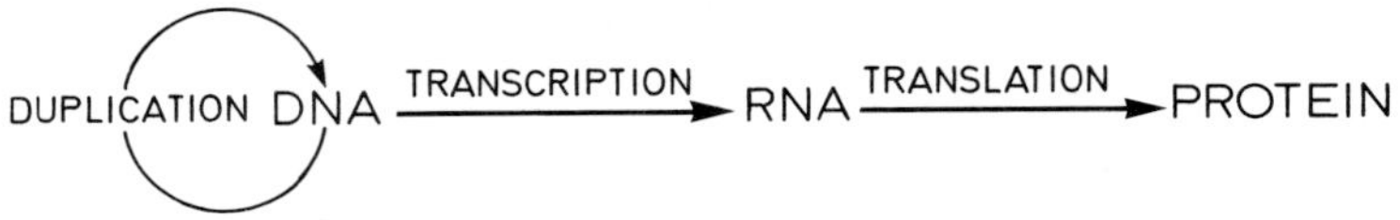

DUPLICATION DNA $\xrightarrow{\text{TRANSCRIPTION}}$ RNA $\xrightarrow{\text{TRANSLATION}}$ PROTEIN

Fig. 4.2 The central dogma.

messenger-RNA template. Since cells are continuously synthesising new proteins there is also a requirement for RNA synthesis throughout the cell cycle whereas DNA synthesis is confined only to a part of the cycle. If the pathway of Figure 4.2 is now expanded then it can be stated that radiation damage to the cell is followed by expression of molecular damage in the following sequence:

DNA——>RNA——>proteins ——>lipids and other macromolecules.

The relationship between intracellular damage to DNA and the response of the cell (lethal or otherwise) to radiation is one of the fundamental problems of molecular radiobiology. Absolute sensitivities of living cells to the lethal effects of radiation vary widely but, in general, mammalian cells are particularly sensitive. In well oxygenated suspensions radiation doses of the order of a few hundred rads suffice to prevent division of the large majority of the cells. From purely radiation chemical considerations, the amount of molecular damage that can arise from the absorption of such small amounts of radiation is extremely small. Most of the large macromolecules such as proteins and enzymes, that are essential to the functioning of the cell, are represented many times in its structure. It would not be expected that damage to a few of these molecules would bring about such a drastic response to radiation. However, there are not many nucleic acids in that structure. The DNA molecule and its sequences of bases is unique. It is not surprising therefore that damage to DNA is presumed to be the main cause of lethality in cells after doses in the clinical radiotherapy range.

Changes in cell population due to selective killing, failure to reproduce, or translocation of certain types of cells are common in many organs after irradiation, e.g. bone marrow, testis, intestine, and lymphoid organs. Since the enzymatic endowment can vary from one cell type to another, a shift in the cell population of an organ in favour of one cell type may result in a change in the biochemical activity of that organ.

Macromolecular synthesis

Biochemical changes in single cells can be due to altered synthesis or catabolism of cellular constituents, or to the loss from their normal intracellular sites in the cell. Moreover, cells can be arrested by irradiation at a certain stage in their cell cycle, e.g. cannot progress from the earlier G_1 phase to S (DNA synthesis), and, therefore, will not produce the enzymes needed to synthesize DNA. Finally, cell death may manifest itself in a variety of biochemical alterations. Only by establishing the time sequence of the different biochemical changes can one avoid confusing cause and effect in such changes in dying cells.

Most of the macromolecules of biochemical importance have a three-dimensional complex structure. Because of this complexity, it is often difficult to recognize or measure chemical changes in these molecules directly. Many of the macromolecules, however, have biological activities or physical-chemical properties which can be quantitated, and which can thus serve as a measure of the structural integrity of the molecule. For example, a decrease in the solubility of a particular macromolecule may suggest an increase in cross-linking; a decrease in the ability of an enzyme system to catalyze a reaction may suggest a chemical change in the reactive group of the enzyme. Moreover, it is often possible to measure functional changes in molecules when they are not a part of a living system. It is not always correct to extrapolate these findings to a living situation where the molecule may be in a different chemical form and may be surrounded by other molecules with differing radiosensitivities and protective capacities.

DNA strand breakage

A particular example of this is the information from the study of single strands of DNA, which can be obtained by lysis of cells under alkaline conditions. High speed centrifugation of such DNA in a sucrose gradient can then be used to determine the molecular weight of the single strands from their sedimentation velocity in the gradient. A significant fall in molecular weight is indicated by a decrease in sedimentation velocity, and this is found to decrease with increasing radiation dose. The molecular weight falls to a minimum within 30 seconds of irradiation and then begins to rise back towards the starting level within 20 minutes. Such studies provide evidence for breakage in single-strands of DNA and the subsequent repair of such breaks. In the living situation, however, DNA is double-stranded (in the well-known double helix) and is associated with protein as a complex nucleoprotein in the chromosomes. It has sometimes been tempting to correlate the laboratory phenomenon of the repair, or rejoining, of breaks in single strands of DNA with the biological phenomenon of recovery from sub-lethal or potentially-lethal damage (considered in Ch. 7). But there are many intermediate steps which must be shown to occur before the biochemical 'repair' of DNA can be equated to 'recovery' from sub-lethal radiation damage. This is a subject for molecular radiobiologists, however, and it is sufficient for the clinician to be aware that considerable research continues into these problems. The only intermediate step in the argument which will be discussed later in this chapter concerns damage to DNA at the chromosomal level.

Biochemical radiosensitisation

Reference was made in Chapter 2 to chemicals which sensitise because they are *electron-affinic* and swing the radiation chemical 'competition' in favour of the oxidative and against the reductive pathway. The chemicals will be useful, if they can be employed at a non-toxic dose level, because they will sensitise hypoxic cells in tumours but will have no effect on normally oxygenated tissues. Such sensitisers might be classified as biophysical in action, in contrast to the more biochemical approach to radiosensitisation which is to use compounds which affect DNA; e.g. by an alteration or delay in DNA synthesis. The halogenated pyrimidines 5-BUdR and 5-IUdR have been tried clinically on this basis. These thymidine analogues do indeed sensitise but only to the extent to which they are incorporated into the DNA (Szybalski, 1974). A reduction in the size of the shoulder of the cell survival curve (Ch. 6) and the reduced recovery from sub-lethal damage (Ch. 7) may explain the increased radiosensitivity with these pyrimidines, but toxic concentrations of the compounds are usually necessary to achieve this. Furthermore there is no difference in their effect between normal and malignant cells. The usual clinical problem of achieving a favourable therapeutic ratio between normal and tumour damage applies to these compounds, therefore, and also to chemicals like 5-fluoro-uracil and Methotrexate which sensitise cells by delaying their progress through the DNA synthetic phase of the cell cycle (see Ch. 5).

Proteins

Proteins are intimately involved with almost all cellular functions. Among the proteins are many of the structural elements of the cell; enzymes which catalyze the essential chemical reactions, many of the hormones which regulate metabolic processes, and the antibodies which are produced to counteract harmful agents. 'Simple' proteins are chains of amino acids; 'conjugated' proteins contain an organic chemical moiety in addition to the amino acids. For example, nucleoproteins contain nucleic acids plus amino acid chains, and glycoproteins contain carbohydrates in addition to the amino acids. The amino group is the most radiosensitive portion of an amino acid. However, in the formation of a protein, this group is linked to a carboxyl group and is not easily removed from the molecule. Similarly, the carboxyl group is no longer available for a reaction. The 'side chains' (R) are the more radiosensitive portions of a protein molecule. The specific changes that occur in the side chains depend on their chemical composition. A hydrogen atom may be removed or a break may occur between almost any of the atoms in the chain.

Loss of function of a protein by irradiation is not usually due to breaking peptide bonds or otherwise disrupting the primary skeletal structure of the peptide chain. It may result from a change in a critical side chain or from a break in the hydrogen or disulfide bonds which maintain the secondary and tertiary structures. Such a break can lead to a partial unfolding of the tightly-coiled peptide chains which, in turn, can result in a disorganization of the internal structure, a distortion of necessary spatial relationships of side chain groups, or an exposure of amino acid groups resulting in a change in chemical activity. The hydrogen bonds of the secondary and tertiary structure are weak bonds. A number of them will be temporarily broken in the vicinity of an ionization because the sudden introduction of a charge disrupts electrical dipoles. One primary ionisation, then, can alter the structure of the molecule and lead to extensive change in the overall chemical reactivity.

Very sensitive tests are required to detect such damage to proteins after radiation doses in the usual therapeutic range. After very high doses a reduction in enzyme synthesis may be evident as a depression in protein synthesis but even this may not produce any noticeable cellular changes and certainly no cell killing, except perhaps in the more sensitive lymphoid tissue. It is because of the continuous synthesis of the numerous enzymes in mammalian cells that the loss of a sizeable fraction of them following a dose of irradiation is usually of little consequence.

Nucleic acids

In contrast to the relative insensitivity of proteins the nucleic acids, RNA and DNA, show more damage after irradiation. A sensitivity ratio of 1000:1 was described earlier when comparing the effects of tritiated thymidine and tritiated water, i.e. much less damage from non-specific irradiation of the whole cell in contrast to irradiation localised specifically to the DNA. If tritiated uridine is used in this sort of experiment then RNA will now be specifically labelled and the ratio of DNA:RNA sensitivity comes down to a factor of 5:1. This rather low ratio is attributable to the fact that RNA synthesis continues throughout the cell cycle and much RNA is situated in the vicinity of the DNA.

In fact there are various forms of RNA within a cell. Messenger RNA (mRNA) moves genetic information from the nucleus to the cytoplasm, while ribosomal RNA (rRNA) from the nucleoli forms the core of the ribosomes in the cytoplasm where transfer RNA (tRNA) assembles the amino acids into the correct sequence for each protein molecule. The tritium labelling experiment can be extended if tritiated amino-acids like histidine or lysine are used in comparison with tritiated uridine or tritiated thymidine. Then the sensitivity of cells to tritium can be compared for the whole sequence of the central dogma (Fig. 4.2) and the following order of sensitivity to tritiated cell components is found:

$$DNA > mRNA > rRNA \text{ and } tRNA > \text{amino-acids}$$

with damage to DNA being 8 times as effective as damage to these particular amino acids. This more sophisticated approach to molecular radiobiology might have allowed the relative sensitivity of the various cellular targets to be compared. The low ratio of 8 is misleading, however, not only because a considerable proportion of the total complement of cellular RNA is closely associated with the DNA, but also because both DNA and RNA are associated with protein moieties as nucleoprotein. Thus the beta radiation emitted from the RNA and amino-acids will contribute to irradiation of the DNA in such comparisons of tritium uptake.

Division delay

In addition to causing cell death, or to be more precise, loss of reproductive capacity, damage to these various targets may lead to delay in progress through the metabolic pathways inherent in the cell cycle. Figure 4.3 shows how such delay varies with the point in the division cycle when the cell is irradiated. In the 20 hour period depicted, there is much less delay at 4 hours than at other times. This 20 hour period corresponds to the total time for a HeLa cell to progress from one mitotic division to the next and at 4 hours the cell will still be in the middle of the G_1 (or first gap) phase before the onset of DNA synthesis (in the S

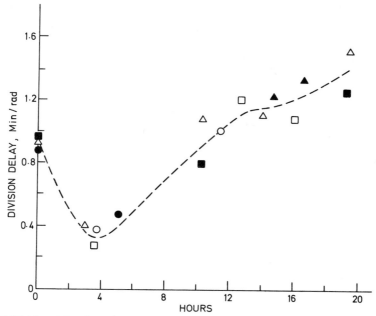

Fig. 4.3 Division delay throughout the HeLa cell cycle. (from Terasima and Tolmach, 1963)

phase). By 8-10 hours the cell will have completed the biochemical processes needed to initiate DNA synthesis and will have entered the S phase. Because DNA synthesis is relatively much more radiosensitive than RNA and protein synthesis there is an increasing amount of delay in progress through the cycle if the cell is irradiated later in the S phase. Depression of DNA synthesis is only a temporary phenomenon after clinical doses (i.e. of a few hundred rads each fraction) but this is still sufficient to interfere with the progress of cells from the S phase into the G_2 (or second gap) phase which is the shortest part of interphase.

An even greater degree of division delay occurs when cells are irradiated in the G_2 phase (i.e. from 16 hours onwards in Fig. 4.4). This is because of inhibition of the synthesis of those proteins necessary for mitosis. Synthesis of these particular proteins is restricted to a very brief period near the end of the G_2 phase. Once this synthesis is completed, however, the cell will have passed a 'point of no return' and irradiation will not then delay further progress towards mitosis. By contrast, cells irradiated earlier in the G_2 phase will be unable to proceed past this critical point if protein synthesis has been inhibited by the radiation. Because the whole G_2 phase is relatively short there is very little time for such cells to repair this damage and overcome the inhibition of protein synthesis. This is the reason why the duration of mitotic delay is maximal for cells irradiated in early G_2 and is less and less evident in cells irradiated earlier in the cell cycle. Damage to cellular targets at earlier points in the cell cycle can often be repaired in a matter of an hour or so and cells are more likely to be restored to an apparently normal state before they reach that critical point

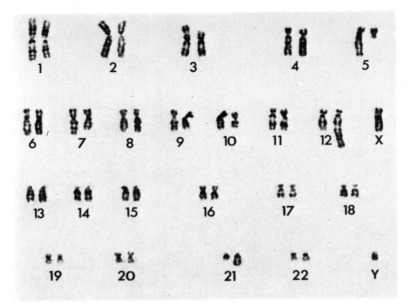

Fig. 4.4 Human chromosomes after radiotherapy. (from Bigger, Savage and Watson, 1972. Fig. 2, chromosoma (Berlin) **39**, 301)

preceding mitosis. Thus irradiation has little effect upon the progress of cells from the G_1 phase into the S phase because there is ample time for the damage to be repaired before the end of the cycle, and division delay remains minimal. Damage to cells irradiated in mitosis is less easily repaired but is manifest towards the end of the subsequent cell cycle in a prolongation of the G_2 phase. Thus depression of DNA synthesis and G_2 block are the main causes of division delay. The relative delay to mitosis varies with the different phases of the cell cycle at the time of irradiation (Fig. 4.3) but it may be taken that on average the delay amounts to about one minute for each rad of exposure. In clinical radiotherapy, therefore, this elongation of the cell cycle is not a very important phenomenon either in the growth restraint of malignant tissues or in retardation of the healing of normal tissues.

Chromosome damage

The more important effect of radiation upon the biological target will be manifest when there is a loss of reproductive capacity rather than a temporary delay in the division cycle. Except for cells in the lymphocyte series and the oocyte which may show death during interphase, the biological target responds to clinical doses of radiation by mitotic death i.e. the cells do not die immediately after irradiation but begin to die when they come into mitosis. At this stage abnormal mitotic figures may be evident as chromosome aberrations or the cell may just fail to divide. Metabolic activities may continue, however, and such cells may remain 'alive' in that they may show no loss of functional integrity. On the other hand the cells are 'dead' for they have lost their capacity for unlimited proliferation, that is, their reproductive integrity. They are 'sterilized'. This is the essential aim of radiotherapy as far as malignant cells are concerned; it is also the unavoidable consequence for many normal cells in the treated area.

Some cells sterilised by radiation may enlarge to 'giant' proportions and become polyploid, i.e. have an increased number of chromosomes, but others may not show any overt chromosomal damage even by the most sophisticated cytogenetic techniques since the damage to the most sensitive target, the DNA, may only need to be at the genetic level. It will be at this level, in fact, that radiation carcinogenesis (discussed in Ch. 12) will operate, since the transformation of normal cells to malignant cells will only be significant if some genetic change occurs which enables such a malignant cell to continue reproducing without the growth control mechanisms of a normal cell. Genetic change of a greater extent will lead to reproductive death which may be manifest as overt chromosomal damage.

If the irradiation is delivered early in the cell cycle before DNA synthesis has begun, then the chromosomes will not yet have duplicated and damage will occur to whole chromosomes. This may be visible at the next mitosis as *chromosome aberrations* as distinct from *chromatid aberrations*. The latter may follow irradiation later on in the cell cycle when each chromosome will have divided into two chromatids held together only at the centromere so that the damage may only be evident in one of the two chromatids of a pair. A large number of permutations and combinations may then be evident in the mitototic figures of an irradiated cell population. This has been the subject of extensive studies by geneticists who have devised a nomenclature to describe the aberrations in terms like breaks, deletions, deficiencies, duplications, interchanges and interchanges, inversions, translocations and other rearrangements (Lea, 1955). These will be the aberrations which remain by the time the cells have reached their next mitosis after being irradiated. During this period of time much of the chromosomal damage may have been repaired although lesions may remain at a genetic level.

Recent advances in *chromosome staining techniques* have revealed much more detail in the form of banding patterns along the length of each individual chromosome. This enables the precise nature of radiation induced chormosome aberrations to be studied and a typical example of bandings is shown along the chromosomes in Figure 4.4. These chromosomes are those from a cell which had been cultured from human skin which had received a dose of 4500 rads nearly 10 years previously. The usual cytogenetic procedure has been followed whereby the cell has been arrested in mitosis so that the chromosomes are in their condensed form and can then be spread out, stained and photographed. The individual chromosomes are then cut out from the photograph and arranged in order of size according to the Denver classification. The figure shows the 22 pairs of autosomes, together with the X and Y chromosomes. It reveals three translocations and one terminal deletion which can be attributed to the radiotherapy. Because of the new staining technique it is possible to say that genetic material has been translocated from chromosomes 3 to 21, 5 to 12 and 7 to 10 and that further material has been deleted altogether from chromosome 7.

Quite clearly the cell whose chromosomes are shown in Figure 4.4 had not received a lethal dose of irradiation since it had survived for ten years and had almost certainly gone through a number of cell cycles during that time, so that these particular aberrations were not lethal. It is difficult to obtain direct evidence of the lethality of chromosome aberrations just because cells so affected will necessarily be absent from the irradiated volume unless special techniques are used to arrest them in the first mitosis after irradiation. Even then one can only determine the proportion of the cells which show aberrations, and indirect estimates have then to be made of the likely outcome of such damage. The lesions may be classified into one or two-hit aberrations depending whether one or two ionising events would be required for their production. With human cells there are so many sites in the nucleus where chromosomes and chromosome arms are close enough together to be damaged by a single ionising event, that the majority of aberrations are of the single hit variety. These effects are *not* dependent upon the dose-rate of the radiation and occur at a rate of 20 aberrations per 100 cells after a dose of 100 rads. The incidence of two-hit aberrations is much lower and, at a typical clinical dose-rate of 100 rad/minute, only about one two-hit aberration will be found in every hundred cells after a dose of 500 rads. This incidence will be decreased if the dose is fractionated because of repair processes.

As far as clinical radiotherapy is concerned the main importance of chromosome aberrations is firstly whether they lead to cell death — which is the object of the exercise as far as the tumour cell population is concerned — and secondly whether they lead to a viable cell population so altered in genetic information as to be deleterious to the human subject. These will be mutations which may result in malignant disease and other changes as a late effect of the radiation. (This subject will be further discussed in Ch. 12) Other long-lived chromosomal aberrations may provide evidence of an absorbed radiation dose which may have been in doubt, (e.g. after a radiation accident) and reliable estimates can now be made of the size of such a dose, by measuring the number and type of chromosomal aberrations in cultured cells.

REFERENCES
Bigger, T. R. L., Savage, J. R. K. & Watson, G. E. (1972) 'A scheme for characteristic ASG banding and an illustration of its use in identifying complex chromosomal rearrangements in irradiated human skin'. *Chromosoma,* **39,** 297-309.
Szybalski, W. (1974) 'X-ray sensitization by halopyrimidines.' *Cancer Chemotherapy Reports,* **58,** 539-557.
Terasima, T. & Tolmach, L. J. (1963) Variations in several responses of HeLa cells to X-irradiation during the division cycle. *Biophysical Journal,* **3,** 11-33.

FURTHER READING
Altman, K. I., Gerber, G. B. & Okada, S. (1970) *Radiation Biochemistry.* New York: Academic Press.
Lea, D. E. (1955) *Actions of Radiations on Living Cells.* Cambridge: University Press.

5. The Cell Cycle

Earlier chapters in this book have sought to identify the biophysical events (Ch. 2) which follow the ionisation of an irradiated tissue and the resultant biochemical damage (Ch. 4). The intervening chapter (Ch. 3) identified the cell as the biological target and the old Law of Bergonié and Tribondeau was quoted to illustrate the concept of *radioresponsiveness*. It was shown that when a tissue is irradiated with doses in the usual clinical range, then the rate of response of the tissue will depend upon the rate of proliferation of cells in that tissue. This is because mitotic death is the usual form of permanent radiation damage so that a cell population with a low mitotic index will be slow to respond to radiotherapy, and vice versa. For this reason it is necessary to know how a human cell population grows in number, and the various ways by which this rate of growth can be measured. The final section of Chapter 3 contained a brief discussion of this topic and Figure 3.4 illustrated a single cell cycle. The period of time between one mitosis and the next is called interphase. This is sub-divided into three phases, because DNA synthesis is confined to a central part of interphase. All four phases of the cell cycle are M (mitosis) then G_1 (the first gap before the beginning of DNA synthesis), S (the period of DNA synthesis) and G_2 (the second gap, after DNA synthesis has been completed and before the cell begins the next mitosis). Knowledge of the duration of these four phases, and of the whole cell cycle, will contribute to an understanding of the effects of radiotherapy on normal tissues.

Cell population kinetics

The word kinetics means the study of motion. Cellular kinetics describes the growth of a cell population, used in relation to the change in total number rather than the circulation of individual cells (where the word 'dynamics' is used). This change in total cell number may show as a net increase, in growing children and tumours, or as a net decrease after cell damage (such as by radiation). The commonest circumstance in an adult patient is for a cell population to remain static in total number although a rapid turnover may continue in the individual cellular components. This turnover of cells, or kinetics of the population, can be measured if suitable methods are used.

Increase in tissue volume. Where there is a net increase in a total cell population the volume of the appropriate tissue can be observed to increase. The commonest example of this is the growth of the embryonic and immature child. Direct measurements of linear parameters (length, breadth, girth) of the whole child, or its individual parts, provide quantitative data on tissue volume. The total volume of a child can also be derived from its weight. The fact that normal growth is not exponential (except in the early embryo), and that cellular differentiation increases in relation to cellular proliferation, makes the kinetics of the growing whole child a complicated subject for study.

The next most important example of a net increase in total cell population is found in tumours. Their cellular kinetics may be no less complicated than in some growing normal tissues but they still permit measurements of total tissue volume. The diameter of an accessible tumour must be one of the commonest of all clinical measurements in cancer work. It must be remembered that a tumour may increase in size due to proliferation of connective tissue or vascular elements, increased blood supply, haemorrhage, oedema or cyst formation; not solely because of growth in the number of tumour cells. Nevertheless, while increase in tumour diameter may provide positive evidence of tumour cell proliferation, the converse may not be true since proliferation may still continue with no apparent change in tumour volume. Nevertheless, since such data may be the only ones obtainable from human tumours they may still provide a useful parameter for measuring the effects of radiations on tumour cell kinetics.

Increase in total cell number. Direct estimates of total cell number allow a much more accurate measurement of the kinetics of a cell population. Where the population is homogeneous, cell counts indicate the rate of growth of such a population. Unfortunately, such homogeneous populations are rarely found except in cell cultures and ascites tumours, although clearly differentiated cell types (like elements in peripheral blood) may be counted with almost the same degree of accuracy. More specialised tissues, like the epithelial lining of the intestinal wall, may also be analysed by counting the total number of epithelial cells in the average crypt and villus.

Analyses of such specialised tissues require a reproducible criterion to enable a quantitative selection of the cells of interest from all other cells in the sample (e.g. selecting small lymphocytes from the remaining cells in the peripheral blood). This criterion is essential if the selected cells are to be considered as a homogeneous population for kinetic studies. Even when this requirement is met, however, data from homogeneous cell populations require care in interpretation.

The simplest expression of an increase in the total number of a homogeneous cell population is the growth curve. This will be exponential so long as the growth conditions remain unchanged. Such a requirement is met by an established cell line in tissue culture. Figure 5.1 is an example of the growth of HeLa cells over a period of 16 days during which the cell population doubled 13 times; amounting to nearly four factors of ten in growth. The cell population doubling time can be calculated from such a curve to be 30 hours but normally such an estimate represents the mean *doubling time* of the cell population and not the mean *cell cycle time.* This is because the doubling time can only be equated to the actual cycle time in the case of a homogeneous cell population, all of which is in exponential growth. The mean doubling time of a cell population will be longer than the mean cell cycle time, whenever there is significant proportion of the cells which are not in the *Growth Fraction* and when there is a significant *Cell Loss Factor* with resultant *Random Death* in the cell population.

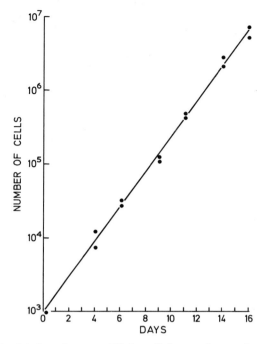

Fig. 5.1 Growth curve of HeLa cells in monolayer culture.

The meaning of these various terms will be explained in the remainder of this chapter, together with a description of some of the special techniques which may be used clinically to measure the kinetics of human cell populations.

Random death

Even a comparatively homogeneous cell population, like the recently recloned HeLa cell line in Figure 5.1, will include a range of cell cycle times which may extend to infinity. This implies the presence in the total population of a proportion of cells which do not divide at all. Although such cells may not be removed from the tissue for a variable period of time, such cells may be said to have suffered *Random Death*. As far as the kinetics of a proliferating cell population are concerned, cells which have effectively stopped dividing are reasonably entitled *dead* even if they are as yet only sterilised and will die later. (The definition of proliferative capacity is discussed in Chapter 6.)

With a fixed proportion of such random death the resultant growth curve will still be exponential in shape. Figure 5.2 illustrates four examples of a theoretical cell population in which the cycle time of the proliferating cells remains constant but an increasing percentage of cell sterilisation lengthens the population doubling time. If these were clinical measurements of total cell number, there would be no way of differentiating this effect of random death from the effect of a simple increase in cell cycle time (with no random death) which would result in a growth curve of the same slope.

If the cell cycle time remains constant but there is increasing random death, the population doubling time will apparently lengthen until, with 50 per cent random death the cell population will appear to be static in size even though half the cells are still proliferating at the normal rate. This is the sort of *steady state* which applies to normal cell populations in the healthy adult, with proliferation exactly equalling cell loss. With 51 per cent or more random death the population will decrease in size. In the context of tumour therapy this would be a satisfactory result, but the observed phenomenon might still conceal a proportion of actively proliferating tumour cells. A growth curve of total cell number may therefore provide a misleading estimate of the kinetics of a cell population, unless this is known to be homogeneous and the proportion of random death has been determined.

This parameter and the parameters for the 'living' proportion of a cell population can all be determined by the various techniques which have become standard in the field of cell population kinetics. Most of them were developed in the laboratory using tissue cultures or animal systems, but many are applicable to the clinical situation. There are two ethical problems which limit the extent of such clinical determinations, however. Serial sampling of the cell population will often require multiple biopsies of the same volume of tissue. Isotopic labelling will usually involve some degree of total body irradiation. Assuming these two objections are overcome then the following techniques, developed in the research laboratory, can be applied clinically.

Pulse labelling of DNA

Apart from the very brief period in the cell cycle occupied by mitosis, the one positively measureable event in the cycle is DNA synthesis in the middle of interphase. A radioactive label can be applied to the cells using the specific DNA precursor, thymidine. The labelled cells can then be studied by high resolution autoradiography which will enable the behaviour of individual cells to be measured but will require sampling and microscopic preparations of the cell population. Alternatively, the extent of labelling can be measured by scintillation counting which will enable a tissue to be studied *in situ* but with no positive identification of which cells are labelled, nor the extent to which they are labelled. Suitable combinations of these two techniques, repeated at intervals during the period of clinical significance, will provide values for most of the kinetic parameters of a cell population.

In the majority of recent studies, ^{3}H-thymidine (tritiated thymidine) has been used as the DNA label, although the problem of the radiobiological effect of prolonged incubation with tritium has led some workers to suggest the preferential use of ^{14}C-thymidine as an alternative when a short duration of labelling is not applicable. More recently, because of the possible pool effects and reutilisation problems with thymidine, ^{125}IUdR (5-iodo-2-deoxyuridine) has been used. This has enabled estimation of cell loss rates from tumours where reutilisation of thymidine is likely to be a significant factor.

In practice, it is more useful to use a labelling period which is short in relation

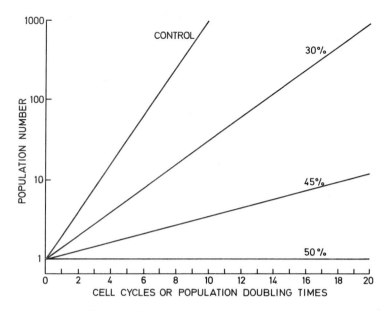

Fig. 5.2 Growth curves with increasing percentages of random death. (from Lajtha and Oliver, 1962).

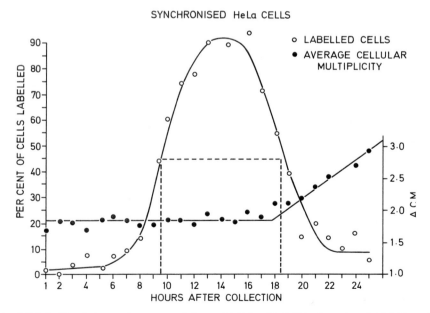

Fig. 5.3 Labelling index and average cellular multiplicity (A.C.M.) of synchronised HeLa cells. (from Nias and Fox, 1971).

to the DNA synthetic period of the cell cycle, i.e. a 'pulse' label. A 10-30 minutes exposure of an asynchronous cell population to ^{3}H-thymidine will label only those cells that are passing through the S period at that time. Samples of the cell population are then fixed at suitable intervals, (e.g. hourly or two hourly) after labelling and three types of analysis can be made from autoradiographs of such pulse-labelled cell populations: percentage labelling, grain count halving and labelled mitoses. The relative duration of the phases of the cell cycle can be estimated to varying degrees of accuracy by the three methods.

Percentage labelling. Certain cell populations in tissue culture can be synchronised (with respect to the phases of the cell cycle) to a relatively high degree. Under these circumstances the percentage labelling curve can be used to confirm the degree of synchrony and to estimate the phases of the cycle. Figure 5.3 shows a curve obtained with HeLa cells synchronised by the mitotic selection method. With this technique the pulse label is applied just before fixation of each sample of cells for autoradiography. The samples are taken at intervals following selection of cells which were in mitosis at zero time.

As the synchronised cells progress through the cell cycle the labelling index rises from nearly zero in the G_1 phase to nearly 100 per cent in the middle of the S phase. The index then falls as the cells progress through the G_2 phase into mitosis. It does not fall to the starting percentage because the degree of synchrony is already reduced by the time the cells reach the second half of the cycle. For this reason, although the duration of the G_1 and S periods can be estimated with some confidence, estimation of the G_2 period and the length of the total cycle require additional data such as the curve of *Average Cellular Multiplicity* (A.C.M.) also shown in Figure 5.3.

This curve remains on a plateau for the greater period and then begins to rise as cells come into mitosis. Such a curve confirms the degree of synchrony of the cell population during the first half of the cycle. This can also be estimated from the height reached by the labelling index as well as from a simple determination of the mitotic index of the cells selected at zero time.

Grain count halving. If the average number of silver grains is determined from a representative sample of labelled cells at various time intervals following the pulse label of a cell population, the label will have been diluted by the subsequent divisions of the cells and the half-time of the decrease in average grain count per labelled cells should provide an estimate of the mean cell cycle time. This can only be used to determine the cell cycle of a population if it can be assumed that any cells which leave the population are a true sample with respect to tritium labelling, so that any changes in the average grain count are not the result of any such cell loss. This requirement may still be met by cells which 'leave' a proliferating population when they 'move' into a non-proliferating maturation 'compartment' before actually leaving the population.

The method has the advantage that, in cell populations where there is a significant proportion either of random death or of non-proliferating cells, it is only the proliferating cells that are assessed. Thus it will be the *mean cycle time* of these cells which will be estimated rather than the mean doubling time of the whole population, which is often a less useful parameter. This advantage must be balanced against the disadvantage of having to count sufficient numbers of grains for statistical accuracy. In addition, there is the problem that if there are to be sufficient numbers of silver grains over the daughter cells for accurate counting after two to three divisions, the original cells must be labelled with such a large amount of radioactive isotope that they may suffer radiation damage.

The same information can be derived from scintillation counting which will also provide a measure of the average amount of radioactivity in a cell population. The method can also be used to estimate the rate of cell loss from tumours or the *Cell Loss Factor*. In this respect ^{125}IUdR is the labelling precursor of choice since, following cell breakdown, it is reutilised to a lesser extent than tritiated thymidine. Figure 5.4 illustrates the method for L1210 leukaemia cells which were labelled *in vitro* with ^{125}IUdR, then exposed to various treatments, including increasing doses of radiation and heat inactivation, before being inoculated into mice. The whole body ^{125}I radioactivity of the mice was monitored at daily intervals by placing the mice in a scintillation counter. Figure 5.4 shows the rapid fall in radioactivity for the mice which received killed cells, a slow fall with untreated cells (until the animals died of leukaemia at the point indicated) and progressively more rapid rates of fall after increasing doses of radiation. The rates of fall in ^{125}I radioactivity indicate the actual disintegration of labelled tumour cells, rather than radiation induced loss of reproductive integrity, and the Cell Loss Factor can be directly calculated.

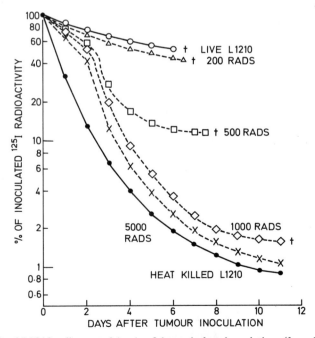

Fig. 5.4 Death of L1210 cells treated *in vitro* 2 hours before inoculation. (from Hofer, 1970).

The method can be used for solid tumours labelled *in vivo*. When the isotope has been injected, studies have shown that 90 per cent of tumour radio-activity is DNA bound from 24 hours after injection so that scintillation counting of whole tumours can then be performed. Values from 0-95 per cent have been found by this method, showing the great variability in cell loss rate between different tumours. Examples of cell loss factors are shown in Table 5.1 in this chapter, and in Tables 10.1 and 10.2. In many human tumours there is such a high cell loss factor that the rate of cell production only slightly exceeds the rate of cell loss and this accounts for their slow growth. In *normal* adult tissues the number of cells remains constant because the rate of cell production equals the rate of cell loss; the cell loss factor is thus 100 per cent.

Labelled mitoses. This third type of analysis of a pulse labelled cell population provides by far the most reliable data on the duration of the cell cycle and its phases. It has the advantage of not depending upon the labelling index (which often varies), nor is it seriously affected by any partial synchrony. The method combines observation of cells at the one fixed point in the cycle which is easily recognisable — mitosis, with the fact that a cohort were in the S phase when the population was pulse labelled. The passage of the cohort of labelled cells can then be followed through successive mitoses and the proportion of labelled metaphases will vary with time, according to curves shown in Figures 5.5 (a) and (b).

From this curve the duration of the mean cell cycle is measured as the time between the same points in the first and second cycles. The average period of S is estimated as the time from the top of the first ascending curve to the bottom of the first descending curve. The duration of G_2 and $\frac{1}{2}$ M is taken as the period from zero time to half way up the first ascending curve. The time for $G_1 + \frac{1}{2}$ M can then be derived by subtraction.

Most of these data must be derived from the first wave of a labelled mitoses curve which is reasonably well defined. Because of variations in the length of the cell cycle this definition is gradually lost, as can be seen already in the second wave of Figure 5.5(b). Even so, this is the best of the pulse labelling techniques for the analysis of the cycle of an asynchronous cell population. As with the grain count halving method, only the proliferating fraction of a cell population will be assessed by the analysis of labelled mitoses. The size of this fraction can be estimated in conjunction with the labelling index. The various methods of assessing the *growth fraction* will be discussed below.

Continuous labelling of DNA. Just as the simple mitotic index can be refined by colchicine treatment of cells to provide a measure of mitotic rate, so the continuous labelling of a cell population represents a refinement of the simple labelling index to provide a labelling rate. Care must be taken to avoid isotope effects from long, continued or repeated labelling, although the label will be continuously diluted in a dividing population and sometimes in other tissues. Repeated labelling may be used instead of continuous isotope incorporation, but the time intervals between administrations of the tracer isotope must not exceed the duration of the DNA synthetic period.

In theory, continuous labelling curves should show some fine structure

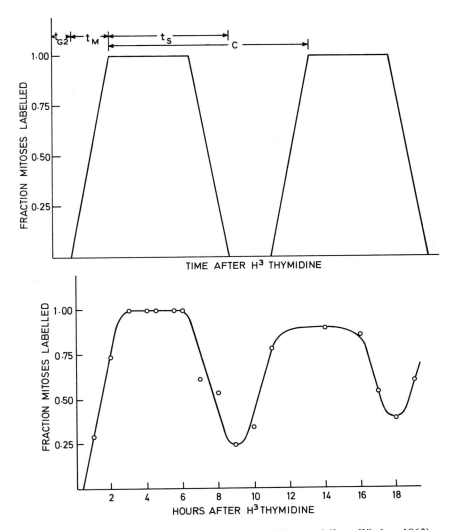

Fig. 5.5 Labelled mitoses curves. (a) Hypothetical; (b) Actual (from Wimber, 1963).

depending on whether steady state or exponentially growing cell populations are being examined. In practice, a fine structure is rarely visible because of the spread of the duration of the phases of the cell cycle. However, the labelling index should reach 100 per cent in time, if the population does not contain a long lived 'non-growth' fraction (see next section). The time taken for the labelling index to rise to 100 per cent is equal to the sum of the G_1 and G_2 periods in an exponential cell population and the percentage of the cycle time occupied by the S phase is indicated by the labelling index at the start of the continuous labelling curve.

This is shown in Figure 5.6 for two populations of HeLa cells. The curve for the control cells (which have a cycle time of 26 hours) shows that 40 per cent of

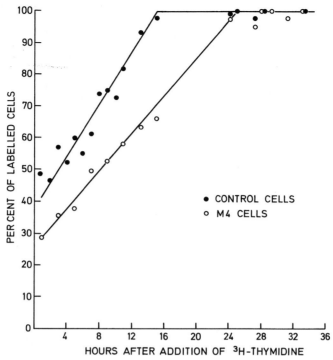

Fig. 5.6 Continuous labelling curves of two populations of HeLa cells. (from Nias, Fox and Fox, 1970).

the cycle is occupied by the S phase and that it takes 16 hours for the cells to pass through the G_2 and G_1 phases, which occupy the remaining 60 per cent of the cycle time. The M4 cells have a longer cycle of time of 34 hours (due to heritable damage from an alkylating agent which caused an elongation of the G_1 phase). The curve for these cells shows that only 30 per cent of the cycle is occupied by the S phase while the G_2 and G_1 phases last 24 hours. In both cell populations the S phase lasts 10 hours, but this information cannot be obtained directly from the curves. Continuous labelling curves can be used to calculate the values of the other kinetic parameters of heterogeneous cell populations both *in vitro* and *in vivo*.

This is, in fact, the best method of obtaining the mean cell cycle time if the spread of cycle times is so great that a second wave of labelled mitoses (in Fig. 5.5(b)) is not seen at all; as is the case with some epithelial tissues. If the labelling index plateaus off at a level significantly below 100 per cent, this level indicates the proportion of cells in the population which are dividing; i.e. the *Growth Fraction,* which will be discussed in this next section.

Tumour cell kinetics

One of the most useful ways of explaining the growth and structure of tumours is the growth fraction model shown in Figure 5.7. In this diagram the

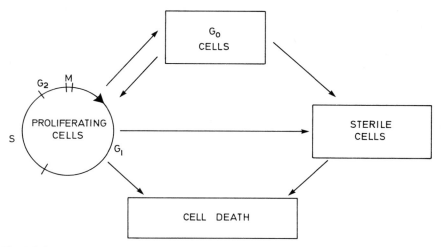

Fig. 5.7 Compartments of growth fraction model. (from Mendelsohn and Dethlefson, 1967).

cell population is divided into four compartments (where 'compartment' describes function, not structure). The proliferating compartment represents the growth fraction (sometimes called 'P', for proliferating cells) from which cells may pass one way into compartments of either sterile or dead cells (cell loss). A reversible pathway exists between the growth fraction and a compartment containing cells in G_0; i.e. resting (or 'Q', for quiescent) cells. These can be recruited into the growth fraction by a suitable stimulus, such as depletion of proliferating cells by radiotherapy. This process of *recruitment* may apply to normal stem cell populations as well as to resting tumour cells. In a simpler form (i.e. without cell death), the model would divide the tumour cells into a proliferating and a non-proliferating pool. The *Growth Fraction* is defined as the ratio of the proportion of proliferating cells to the total cell population. The growth fraction can be estimated from the disparity between the doubling time of the tumour and the generation time of the proliferating cells. In the presence of cell death, the growth fraction model becomes more complex and, after irradiation, changes in growth fraction, cell loss and intermitotic time are all likely to play some part in determining changes in growth rate during the repopulation process. Of these, it is considered that variation in growth fraction and rate of cell loss are the major factors and these in turn are likely to depend on changes in vascular architecture and function after irradiation.

The clinical use of measurements of tumour diameter may suggest that tumours grow logarithmically. By contrast Figure 5.8 illustrates the non-exponential growth of a transplanted rhabdomyosarcoma where tumour volume is the parameter of repopulation. It is seen that with such serial measurements the growth curve does not remain exponential but bends towards the time axis. However, growth curves obtained from serial measurements of tumour diameter frequently do remain exponential during a relatively short period of observation. This paradox has been explained on the basis of central necrosis of a tumour with continued proliferation of the outer cell

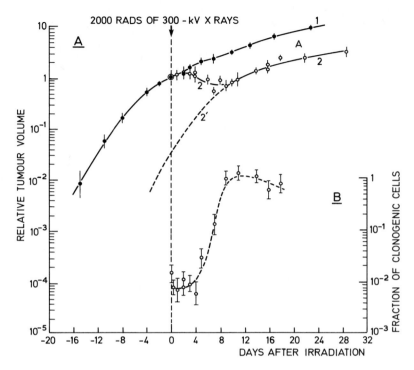

Fig. 5.8 Growth curves of a transplanted rhabdomyosarcoma. (from Hermens and Barendsen, 1969).

layers. These two processes acting together will often result in a linear increase of tumour diameter with time when a thin proliferating shell surrounds a necrotic core — the so-called 'orange skin' effect.

Central necrosis is not the only factor confusing the evidence from serial measurements of tumour diameter. Every tumour contains a stroma of normal tissue and the relative proportions of stromal cells to tumour cells varies in different tumours and at different times during the growth of the same tumour. For this reason serial estimates of tumour volume, whether obtained from total weight or merely from diameter, provide only an indirect measure of the kinetics of the total tumour cell population.

The reason for the decrease in the rate of growth of a tumour with increasing size is probably because the supporting stroma cannot maintain a rate of growth equal to that of the parenchymal cells. The nutritional environment of the tumour cells becomes poorer as the distance between blood vessels increases, and this leads to a decreased rate of cell proliferation and cell death. The rate of proliferation of endothelial cells may also limit, indirectly, the rate of tumour growth but there have been few attempts to study the population kinetics of vascular endothelial cells and supporting connective tissue, possibly because endothelial cells of blood vessels are difficult to recognize in thin tumour sections. Values have been estimated for kinetic parameters for endothelial cells of capillary walls and for carcinoma cells of a transplanted C_3H mouse mammary tumour. The turnover time was about 50 hours for the endothelial cells compared with 22 hours for the carcinoma cells.

It should be noted that the *turnover time* is not the same as the cell cycle time. Nor can values for the kinetics of mouse cells be used to guide radiotherapists in the timing of the response of human cells. If the 2-year life-span of a laboratory mouse is compared with the 70 years expected of man then a considerable difference might be expected in kinetic values. This is not borne out of practice and the two mammalian species show relatively small differences when precise measurements have been possible. The values for colonic epithelial cells provide a typical example: S phase in mouse $6\frac{1}{2}$ hours, in man 20 hours; G_2 phase in mouse $1\frac{1}{2}$ hours, in man 8 hours; cell cycle in mouse 16 hours, in man 45 hours. The human values are included in Table 5.1 where seven cell types are listed. Some values for *Growth Fraction* and *Cell Loss Factor* are included and those for Basal Cell Carcinoma provide a numerical explanation for the very slow clinical progress of rodent ulcers; both in their usually long history of growth before diagnosis and in their sometimes slow disappearance after radiotherapy. This is because of the very high cell loss factor (95 per cent) and the low growth fraction (30 per cent) of this cell population, apart from its relatively long cell cycle.

Table 5.1 Kinetics of some human cell populations

Cell Type	Length of S phase (hrs)	Cell Cycle Time (hrs)	Growth Fraction %	Cell Loss Factor %
Basal Cell Carcinoma	19	72	30	95
Squamous Cell Carcinoma	12	38	24	90
Melanoma	25	80	25	70
Carcinoma Cervix	10	15	50	
Acute Leukaemia	10	20		
Bone Marrow	13	24		
Normal Colon	20	45		

For most human cell populations such precise estimates of kinetic parameters are not easily obtainable. Functional parameters like *turnover* and *transit time* are sometimes used to describe the period from stem cell to functional cell for normal populations like the bone marrow and intestinal epithelium; two of the most important tissues which limit radiotherapy (see Ch. 8). Thus the transit time for gut cells is $7\frac{1}{2}$ days and for the three bone marrow cell types is 8 to 13 days for leucocytes, 4 to 10 days for platelets and 4 to 7 days for erythrocytes. Having said that, however, it must be remembered that the life-span of these three cell types in the peripheral blood is about 10 days for leucocytes and platelets but nearer 120 days for erythrocytes. Thus the leucocyte and platelet cell populations are most quickly depleted in the peripheral blood by irradiation of the bone marrow stem cells but gut damage is manifest even sooner. (These time factors will be discussed with reference to the Acute Radiation Syndrome, in Chapter 9.)

For tumour cell populations the functional parameter is the actual doubling time of growth measured clinically as diameter or preferably as volume. But radiotherapy schedules should more logically be considered in relation to the cycle time of the tumour cells which will be very much shorter than the volume doubling time because of low *growth fractions* and high *cell loss factors*. Tubiana (1971) showed how these various parameters can be measured for human tumours. Thus, for one particular basal cell carinoma the cell cycle time was 3 days, but because the growth fraction was 30 per cent and the cell loss factor was 97 per cent, the clinical doubling time was 10 months.

Detailed measurements of such parameters are not yet possible for most human tumours but an alternative kinetic parameter can be measured from a single biopsy specimen, if this is incubated with tritiated thymidine in the laboratory. The labelling index of such a specimen is then used to calculate a parameter: the *Potential Doubling Time* (PDT) of the tumour. This represents the mean rate of cell production expected in the absence of cell loss and is based upon the mean cell cycle time and the growth fraction. Thus, for the basal cell carcinoma mentioned above, the PDT was calculated as 10 days which, while longer than the cell cycle time of 3 days, is considerably shorter than the clinical doubling time of 10 months. In other types of human tumour the potential doubling times have been found to vary from 1½ days for the Burkitt's lymphoma to 43 days for breast carcinoma. Unfortunately it is still very rare for such observations to be well substantiated. Assumptions have to be made in the calculation, for example, that the S phase is constant for all human tumours, which is certainly not the case.

Many of the techniques described in this chapter can be employed to improve upon the simple measurement of PDT, especially when a series of observations is clinically acceptable upon the same tumour. In such cases the cycle time of the cells can be determined in both the tumour and the limiting normal tissues, together with the other kinetic parameters. In the absence of such precise information, estimates of Potential Doubling Time can provide information on the relative growth rates of different human tumours which may be useful in the choice of fractionation schedules in radiotherapy.

Synchronisation

If the cycle time of the tumour cells could be easily and reliably measured and if some more radiosensitive parts of the cell cycle could be identified, then fractionated radiotherapy might be delivered at such sensitive times. Unfortunately this would require all the tumour cells (which were in the growth fraction) to be progressing through the sensitive parts of the cycle at the same time. The cell population would need to be synchronised. In the laboratory this can quite easily be achieved and Figure 5.3 illustrated this for human carcinoma cervix cells cultured in vitro. In man (and also in mouse) the degree of synchronisation achieved by existing methods has been rather disappointing, to date, but the effect of natural circadian rhythms should not be ignored (see Ch. 13, Fig. 13.8).

A quantitative definition of the degree of synchronisation of a cell population assigns values from 0 per cent for cell populations in logarithmic growth, to 100 per cent for populations which are perfectly synchronised. On this basis, any growth pattern which differs from that obtained in logarithmic growth can be said to indicate some degree of synchronisation. A stricter definition of synchronisation requires that the cell population will double in a time interval which is short relative to the total cycle time. Terms like parasynchronous growth and partial synchrony have been used to describe a small change from the logarithmic growth pattern which nevertheless indicates a low degree of synchronisation. For the majority of biological studies involving comparisons between cells in different phases of the cell cycle, a high degree of synchronisation is necessary before useful data can be obtained.

In clinical practice mitotic selection (used for Fig. 5.3) and other physical methods of synchronisation are not applicable. On the other hand, certain cancer chemotherapeutic agents such as methotrexate, 5-FU and hydroxyurea have the chemical effect of holding the cell population up at the beginning of the S phase and this is particularly useful because that point in the cell cycle is one of the most radiosensitive (see Ch. 6). The timing of the doses of chemotherapy has to be adjusted so as to select a suitable fraction of the tumour cell population which is to be irradiated. If a population of cells growing asynchronously is exposed to a toxic agent for a suitable period of time, then a relatively small cohort of cells will escape and this cohort will be confined to cells in one phase of the cycle. The method is illustrated in Figure 5.9 redrawn

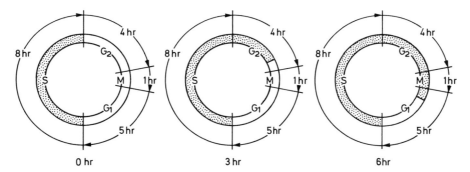

Fig. 5.9 Window method of synchronising cells. (from Whitmore and Gulyas, 1966).

from Whitmore and Gulyas (1966). In this example, high specific activity ^{3}H-TdR is used to destroy the proliferative capacity of all those cells which have passed through the S phase. After 6 hours, only a small cohort of viable cells is left in the remaining 'window' which is limited to the G_1 phase of the cycle. But the window is 4 hr wide and the viable population can only be partially synchronised. The higher the degree of synchronisation that is desired, the narrower the window must be and the lower will be the yield of synchronised cells. This principle of quality, not quantity, has been discussed earlier and is a real limiting factor in the clinical application of these techniques.

There is also the lack of selectivity of action between different cell populations *in vivo*. This is a disadvantage most commonly faced in cancer therapy where the search is for a therapeutic ratio between a tumour cell population which is affected by the chemical and a normal cell population which is not. There is no evidence that any of the chemical agents which have been used for synchronisation *in vitro* have any such selective action *in vivo*, except as a result of differences in cell population kinetics found in the body. Proliferating cell populations will tend to be more affected than resting populations and this has been shown with agents like hydroxyurea. In current terminology this partial synchronisation is described as *Redistribution* of cells within the division cycle and such redistribution can be observed after radiation as well as drug therapy.

REFERENCES

Hermens, A. F. & Barendsen, G. W. (1969) Changes of cell proliferation characteristics in a rat rhabdomyosarcoma before and after X-irradiation. *European Journal of Cancer*, **5**, 173-189.

Hofer, K. G. (1970) Radiation effects on death and migration of tumour cells in mice. *Radiation Research*, **43**, 663-678.

Lajtha, L. G. & Oliver, R. (1962) Cell population kinetics following different regimes of irradiation. *British Journal of Radiology*, **35**, 131-140.

Mendelsohn, M. L. & Dethlefsen, L. A. (1967) Tumor growth and Cellular Kinetics in *The Proliferation and Spread of Neoplastic Cells*. Houston: M.D. Anderson Hospital, p.200.

Nias, A. H. W., Fox, M. & Fox, B. W. (1970) Kinetics of a drug sensitive clone of HeLa cells. *Cell and Tissue Kinetics*, **3**, 207-215.

Nias, A. H. W. & Fox, M. (1971) Synchronisation of mammalian cells with respect to the mitotic cycle. *Cell and Tissue Kinetics*, **4**, 375-398.

Tubiana, M. (1971) The kinetics of tumour cell proliferation and radiotherapy. *British Journal of Radiology*, **44**, 325-347.

Whitmore, G. F. & Gulyas, S. (1966) Synchronisation of mammalian cells with tritiated thymidine. *Science*, **151**, 691-694.

Wimber, D. E. (1963) in *Cell Proliferation* Oxford: Blackwell, p.5. eds. L. F. Lamerton and R. J. M. Fry.

FURTHER READING

Baserga, R. (1971) *The Cell Cycle and Cancer*. New York: Marcel Dekker, Inc.

6. Cell Survival Curves

The question of tumour radiosensitivity is still beset by the apparent contradiction that clinicians find a great variation in radiation response between tumours of different types, while laboratory workers find only relatively small differences in the response of the range of cell types represented in tumours. Standard textbooks of the practice of radiotherapy (e.g. Paterson, 1963) usually contain a classification which divides human tumours into sensitive, less sensitive and resistant classes with respect to the response which may be expected from radiotherapy given to the limit of normal tissue tolerance dosage. This is an operational classification of radioresponsiveness in the sense that resistant tumours are those in which a low therapeutic ratio is found with respect to the normal tissue relevant to such tumours. While the classification is empirical it is based upon many years of observation by experienced radiotherapists. By contrast, results of the more recent studies of radiobiologists show a relatively small range of values from tests of radiosensitivity at the cellular level.

Reproductive death

Cell survival curves are used to show the response of single cells to increasing single doses of radiation of various qualities, delivered under various environmental conditions and at various rates of dose. The response of single cells is tested by their ability to grow into colonies (i.e. to proliferate) after the treatment. If a cell demonstrates this reproductive capacity during a minimum period after the treatment then that cell can be considered to have survived. In this context, survival requires not merely the continued existence of the cell as a living entity (i.e. functionally intact) but also the property of proliferation. When considering the radiotherapy of cancer cell populations, it is that property which is of greatest interest since the aim of treatment is to destroy the proliferative capacity of the cancer cells while leaving sufficient normal cells to maintain the function of the normal tissues in the treated volume. Some of the cancer cells may still 'survive', in the sense that they respire, synthesize macromolecules and exhibit other biochemical properties. In one sense these cells are still living but since they will play no further part in the growth of the tumour they can usually be ignored. This is why the information obtainable from cell survival curves is confined to the response of *proliferating cells.*

The ability of single mammalian cells to proliferate into colonies, was used by Puck and Marcus (1956) in a plating technique analogous to that used for the assay of bacterial cell viability. Single cells were seeded into culture vessels and allowed time to demonstrate their reproductive capacity. Each cell which survived in this sense grew into a colony or 'clone'. The number of colonies containing more than 50 cells (i.e. more than 5 cell divisions) was then counted and, in the vessels seeded with unirradiated cells, there should have been as many colonies at the end of the culture period as there were single cells at the beginning. If different aliquots of these cells were exposed to single doses of irradiation at the time of plating, a dose response was found whereby increasing doses of irradiation damaged an increasing proportion of the cells so that there were correspondingly fewer colonies in those culture vessels. In other words, as the dose of irradiation was increased there was a corresponding decrease in the probability of a cell being able to survive that dose, i.e. an increase in reproductive death.

Dose response curves in vitro

Compared to the number of colonies in an unirradiated aliquot of cells (defined as 100 per cent or unity) the number of colonies formed in an irradiated aliquot will represent the fraction of the initial cell number which survived that dose of irradiation. If values for surviving fraction are plotted against values for radiation, dose response curves such as those in Figure 6.1 will be obtainable. The shape of these curves is characteristic for low LET radiation, i.e. it is the sort of dose response that is found with conventional radiotherapy. After an initial *shoulder* region the larger the dose — on a linear scale — the smaller the surviving fraction — on a logarithmic scale. This *exponential* relationship has been found for all mammalian tissues where it has been possible to test the radiation response of the constituent cells by some quantitative method. What this means is that the larger the tumour volume the

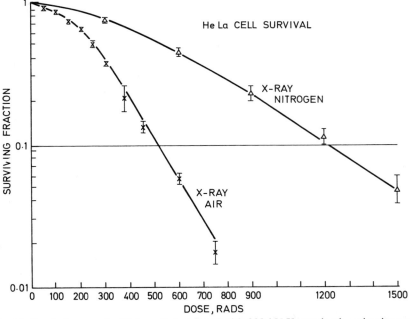

Fig. 6.1 Survival curves for HeLa cells irradiated with 300 kV X-rays in air or in nitrogen.

larger the radiation dose that will be needed to eradicate it — assuming that every single tumour cell must be 'sterilized'. The fact that the normal tissue tolerance dose is smaller for larger volumes is an unfortunate limitation in view of this theroretical requirement for a larger tumour dose. Be that as it may, the essential feature of mammalian cell survival curves is their *exponential* shape following the initial *shoulder* region.

This conforms most obviously with that alternative of target theory (Ch. 2) which assumes multiple targets in the cell each of which must be damaged by a single hit before that cell is lethally damaged, although the multiple hit in one target alternative cannot be excluded. There is one essential difference between experimental cell survival curves and the *multi-target single-hit* theory. This is that the shoulder region is significantly steeper than theory suggests, i.e. a definite proportion of damage occurs even at the lowest doses.

This means that radiobiologists have to use a compromise formula to calculate the parameters of survival curves; combining that of the classical multi-target model with an additional one-hit component. Since individual doses of fractionated radiotherapy lie more frequently over the range of the initial shoulders of survival curves than their final exponential portions, the validity of such formulae has been examined in considerable detail (see Alper, 1975).

The curves in Figure 6.1 illustrate the response of cells to *single* doses of radiation. The response of cells in a tissue to a dose fractionated over several weeks will depend upon a number of factors (Ch. 13) including the size of the individual fraction doses. The biological effect of these doses can be predicted from the cell survival curves obtained with those cell populations which are relevant to the clinical problem. The usual method of comparison between such curves is by describing their shape using the two parameters D_o and N. D_o describes the slope of the *exponential* portion of the curve after the initial shoulder and it is the dose, in rads, required to reduce the surviving fraction to a value of $\frac{1}{e}$ (where e is the exponential function and $\frac{1}{e}$ equals 0.37). Thus D_o is the *mean lethal dose* for that cell population and the value can be read off the graph as the extra dose required to reduce survival from 10 per cent to 3.7 per cent or 1 per cent to 0.37 per cent. The dose required to reduce survival from 100 per cent to 37 per cent might be called D_{37} but this is a misleading parameter since it includes the shoulder portion of the curve.

The size of the *shoulder* of a cell survival curve is described by extrapolating the exponential portion upwards to the vertical axis of the graph. This point on that logarithmic scale is then called, quite simply, the *extrapolation number*, N. The point where this extrapolated line crosses the horizontal axis (at 100 per cent survival) may be described as the *quasi-threshold dose* D_q (in rads). This may loosely be considered as an amount of 'wasted' radiation (attributable to sub-lethal damage, Ch. 7) for that cell population, after a large dose has been given. The shapes of cell survival curves can thus be compared using the parameter D_o to describe the exponential slope and either of the parameters N or D_q to describe the extent of the shoulder. The size of the shoulder itself determines the response to the multiple small doses of about 150-300 rads each, which are commonly used in radiotherapy.

Oxygen enhancement ratio

The difference between the exponential slopes of the two curves in Figure 6.1 illustrates a well known radiobiological phenomenon; namely, enhancement of the effect of radiation by oxygen. If the two curves can be fitted to the same shape, i.e. their exponential slopes can be shown to extrapolate to the same point on the vertical axis (the same extrapolation number, N) then this indicates that oxygen is a purely dose modifying factor in the environment of the cells at the time of irradiation. In such a case, the ratio of the two values for D_o provides a single value for the parameter called the oxygen enhancement ratio (OER). This is the ratio of doses given hypoxically or in air to produce a given level of cell killing. If, on the other hand, pairs of survival curves of cells irradiated in air and under hypoxic conditions can not be fitted to the same shape, i.e. do *not* have the same value for N, then oxygen can no longer be considered to have the same dose modifying effect at all dose-levels and calculations based upon the ratios of the D_o values may provide a misleading indication of the OERs applicable to those cells. In the case of Figure 6.1 the value for N turned out to be 3.22 ± 0.23 and a statistical analysis showed that it was justifiable to use a common shape for the survival curves. Thus, oxygen was a purely dose modifying factor, under the experimental conditions described and the OER obtainable from the ratio of the D_o values of 150 rads for aerated cells and 360 rads for hypoxic cells was 2.39. Calculations of this sort can be applied to data obtained with any cell population which can be assayed for survival of the colony forming ability of single cells.

The data in Figure 6.1 were obtained using HeLa cells which were originally derived from a human carcinoma of cervix. They have since been grown in monolayer culture in a number of laboratories and provide a suitable model for

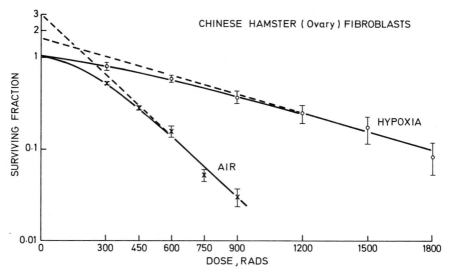

Fig. 6.2 Survival curves for Chinese hamster (ovary) fibroblasts irradiated with 300 kV X-rays in air or under hypoxic conditions.

radiobiological studies of single epithelial cells. These data can be compared with those with the fibroblastic cell line obtained from the ovary of the Chinese hamster, designated the CHO line. Figure 6.2 shows a pair of survival curves found with these cells and this figure has been used to illustrate extrapolation of the exponential slope back to the vertical axis to obtain the size of N, the *extrapolation number.* It is obvious from inspection that different extrapolation numbers apply to the two survival curves in Figure 6.2. The computed values were 3.00 ± 0.47 for aerated cells and 1.67 ± 0.11 for hypoxic cells. These values are significantly different (P ≃ 0.01) and so the ratio of values for D_o would give a value for OER which would be misleading, if applied to small doses in the shoulder region. OER values then have to be obtained from the ratio of doses required to produce a given level of survival and will vary with the level chosen. Despite this, the ratio of D_o values has been used in the last column of Table 6.1 where the parameters of some of the survival curves illustrated in this chapter are listed both for cultured cells, with which hypoxic conditions are easily examined, and for tissue systems in the animal.

A much longer list of the parameters of low LET radiation survival curves for aerated cells is provided in Table 6.4 at the end of this chapter. These are for all the cell populations discussed throughout this book. The table provides some basis for comparison between the radiosensitivity of different types of cell assayed either in cultures in *vitro* or in tissues *in vivo.* Not all the assay systems are strictly comparable, however, and the appropriate pages in the book should be consulted for the details. The very next survival curve to be described, for example, uses an *in vivo* assay method which is quite different from the *in vitro* culture technique used for the survival curves described so far.

Dose response curves in vivo

Spleen colony assay. One of the most radiosensitive survival curves yet measured is shown in Figure 6.3. This curve is included to illustrate two principles: firstly, that leukaemic (and normal haemopoietic) cells are the most sensitive class of cells; secondly, that survival curve data are obtainable for cells assayed *in vivo* as well as *in vitro.* The data in Figure 6.3 were obtained using L1210 leukaemic cells, obtained from the femoral bone marrow of leukaemic DBA/2 mice irradiated with the doses as indicated. Leukaemic marrow cells were then injected into recipient mice and these were sacrificed after seven days and their spleens examined for leukaemic colonies. The assay system is in most respects analogous to that used for cells plated *in vitro* but the irradiation of cells and their subsequent spleen colony formation is all undertaken *in vivo* — only the femoral marrow cells are counted *in vitro* to determine the correct number for injection into recipient mice. This is the method used by Till and McCulloch (1961) for normal haemopoietic cells and adapted by Bush and Bruce (1964) for leukaemic cells. It can be concluded that the same shape of survival curve is found by the *in vivo* method as is found *in vitro* and that *in vitro* data are quite relevant to radiobiological problems which must ultimately be tested *in vivo.*

Another conclusion from the spleen colony method is that similar survival curves are found for normal bone marrow so that there is no difference in

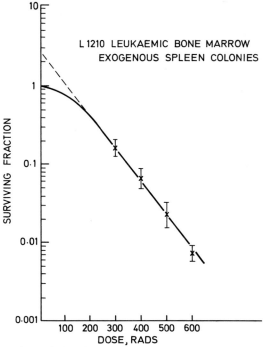

Fig. 6.3 Survival curve for L1210 leukaemic bone marrow cells irradiated in DBA/2 mice with [137]Caesium γ-rays.

radiosensitivity between the normal and malignant bone marrow. If the radiotherapist is looking for an increased therapeutic ratio between the response of malignant and normal tissues, he is unlikely to find it in any difference in intrinsic radiosensitivity of cells of the same histological group. While there are differences between histological groups (e.g. the D_o value of the fibroblasts in Table 6.1 is double that of the leukaemic marrow) even these will provide a limited difference over the fraction dose levels customarily used in radiotherapy. On the other hand, it is not often possible to make the same precise comparison between normal and malignant cells of the same histolog-

Table 6.1 Parameters of cell survival curves.

Cell Type	Aerated cells			Hypoxic cells			Aerated/Hypoxic D_o ratios
	D_o (rads)	N	D_q (rads)	D_o (rads)	N	D_q (rads)	
HeLa (with hypoxia)	150	3.2	160	360	3.2	400	2.39
HeLa (with anoxia)	140	3.1	160	410	1.0	0	2.93
Hamster fibroblast	200	3.0	210	640	1.7	320	3.20
Mouse leukaemia	100	3.0	115				
Mouse intestine	130	—	450				

ical type using the same assay, as it is with the spleen colony method for bone marrow stem cells. Conclusions as to the relative radiosensitivity of different cell types must therefore remain indirect.

Intestinal crypt assay. Another technique for measuring the radiosensitivity of cells in their normal environment is used for intestinal epithelial cells. The cell population orginates in the crypts of Lieberkuhn and if a large enough dose of radiation is absorbed then many of the crypts are left with only about one viable cell. In mouse experiments such cells repopulate the crypts very quickly (the cell cycle time is less than 12 hours) so that after three or four days a histological section across the lumen of the irradiated intestine will show a number of regenerating crypts around the circumference. This number is dose dependent and a survival curve can be drawn as in Figure 6.4. The D_0 of this

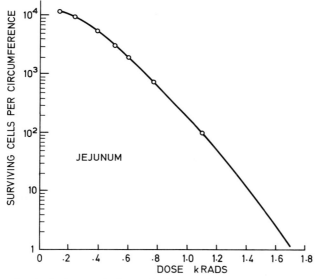

Fig. 6.4 Composite single-dose survival curve for intestinal stem cells irradiated in C_3H mice (from Withers *et al.*, 1975).

curve is 130 rads over the dose range above 1150 rads but it is difficult to derive values for the shoulder region of the curve at lower doses. This is because it is estimated that the total number of stem cells may amount to more than 10 000 per circumference but it is not possible to count more than about 100 regenerating crypts. Over the dose range below 1100 rads the shape of the cell survival curve can only be determined from multi-fraction experiments (Withers *et al.*, 1975). The very large 'shoulder' region on the curve might suggest a large value for N and D_q i.e. a large capacity for the cells to recover from sub-lethal damage (see Ch. 7). The fractionation experiments show that this depends upon the size of the fraction dose, however. With a typical clinical fraction dose of 200 rads, for example, the D_q is 120 rads but this value rises to 450 rads over the high dose range above 1150 rads, where the survival curve adopts its final exponential slope. (These final values D_0 = 130 rads, D_q = 450 rads, have been included in Table 6.1, but see also Fig. 11.4).

Skin epithelial cell assay. This problem of determining the size of the shoulder of a survival curve applies to another limiting normal tissue — the skin. The technique for assaying its radiation response in cellular terms, in its normal environment *in vivo,* will be described in Chapter 8 (Responses of Normal Tissues). A survival curve can be derived, however, and the value for D_o is 135 rads which is within the range of all the other values mentioned in this chapter for aerated cells.

Spheroid cultures

Having described these three examples of survival curves derived from the assay of tissues *in vivo,* and before discussing more cell survival curves derived from monolayer cultures *in vitro,* mention should be made of an assay technique which is intermediate between these two. If cells are cultured *in vitro* but kept in suspension by a magnetic stirrer, they develop into multicellular 'spheroids' which are morphologically similar to nodules of many carcinomas and grow at a comparable rate. This technique is particularly useful for the study of *electron-affinic* drugs like the nitroimidazoles where a quantitative assay is needed not only of the degree to which hypoxic cells can be made more radiosensitive, but also the extent to which the drug can diffuse into the necrotic centre of a tumour. The multicellular structure of these spheroids presents the same sort of diffusion problem for oxygen and other nutrients as is found in tumour tissues *in vivo.* Biphasic survival curves (see Fig. 6.6, later) can be studied under more controlled conditions *in vitro.*

The technique has demonstrated at least one of the consequences of intercellular contact that seems to apply *in vivo.* Thus, some of the survival curves for aerated cells show an increase in the size of the shoulder as the cultures grow from single cells into spheroids of increased diameter. For Chinese hamster (V79) cultures, for example, the extrapolation number is increased from 10 to 100 (with the same D_o value of 170 rads). This is consistent with one of the conclusions that may be drawn from Table 6.4; namely that values for N (and D_q) tend to be higher for survival curves derived from tissue than from single cell assays.

The oxygen effect at low doses

Figure 6.5 shows the pair of survival curves listed in Table 6.1 under HeLa cells (with anoxia). The curve for aerated cells is very similar in shape to that shown earlier in Figure 6.1, although it spans an extra decade in survival because it extends over a wider range of single doses. The curve for 'hypoxic' cells, on the other hand, is quite different in shape from the 'Nitrogen' curve in Figure 6.1. Also extending over a wider range of dosage, it shows a purely exponential shape with an extrapolation number of unity. While the nitrogen curve in Figure 6.1 could be fitted to the same shape as its corresponding aerated curve, the two curves in Figure 6.5 could not ($P \simeq 0.01$). The difference in shape between the hypoxic curves in Figures 6.1 and 6.5 can be attributed to the fact that the hypoxic environment of the cells in Figure 6.1 had been obtained by gassing with white spot nitrogen which reduced the oxygen tension to a low level (<10 parts per million) but not to complete anoxia. The data

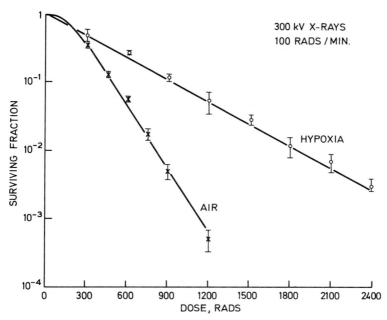

Fig. 6.5 Survival curves for HeLa cells irradiated with 300 kV X-rays in air or under anoxic conditions.

shown in Figure 6.5 were obtained with cells in a crowded cell suspension which respired itself to a lower oxygen tension which can be considered to approach complete anoxia.

The survival data in Figure 6.5 have been listed in Table 6.2 where values for OER are shown at different survival levels. The OER is assumed to be 1 at a survival level of 60 per cent and above. At lower survival levels, the value of OER rises progressively and the ultimate value would reach 2.93 at doses so high that the contribution of the D_q of the aerated curve becomes negligible, so that the aerated/hypoxic D_o ratio is finally applicable. The important conclusion to be drawn from Table 6.2 is that the OER may be considerably lower than this maximum, over the range of single doses customarily used for fractionated radiotherapy. On the other hand, these data apply to the extreme

Table 6.2 Variation of OER with survival level.

Survival Level (per cent)	Dose in air (rads)	Dose under hypoxia (rads)	OER
60	200	200	1.00
45	250	325	1.30
35	300	425	1.42
20	400	625	1.56
10	500	950	1.90
3	650	1450	2.23
1	800	1900	2.37
0.3	1000	2400	2.40

case where the cells are completely anoxic. Such cells may not be of importance in clinical radiotherapy, since they will not retain viability unless reoxygenated within a short period. Such reoxygenation may not be so uncommon, however. By contrast, cells which are hypoxic, rather than completely anoxic, may not show the lower values of OER shown in Table 6.2. It seems likely, however, that a single value for OER cannot be applied to all tumour cell populations over the range of fraction doses used in radiotherapy.

The two curves in Figure 6.5 show the radiation dose-response for cells at the two extremes of complete aeration and complete anoxia. In many tumour tissues, however, the cell population will be some mixture of those two extremes and the actual dose-response curve will be biphasic. Figure 6.6 illustrates this for a mouse tumour studied by the dilution assay technique of Hewitt and Wilson (1959). Leukaemic cells infiltrate the liver of the donor mouse and a cell suspension is then made. After increasing doses of radiation to the donor mice, increasing numbers of cells must be inoculated into the groups of recipient mice if tumours are to develop. By suitably varying the cell dilution, an average of one viable cell is inoculated into each mouse and the usual exponential dose-response is established. If the donor animal is alive at the time of irradiation then its tumour cells are oxygenated and the lowest curve is obtained. If the animal is killed long enough beforehand, all the cells will be anoxic and the uppermost curve is found. With intermediate time intervals, the other curves will be obtained which are biphasic; starting with an aerated slope and changing to an anoxic slope when all the aerated cells have been sterilised.

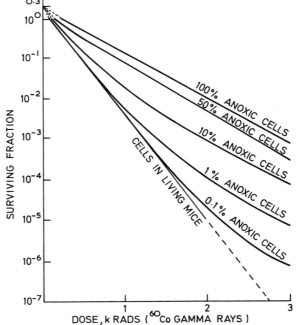

Fig. 6.6 Diagram of survival curves for cell population with various proportions of anoxic cells (from Hewitt and Wilson, 1959).

The relevant curve for radiotherapy is probably the 10 per cent anoxic cell example, although information on the degree of hypoxia of human tumours is very inadequate and a broad range of values can be presumed to apply. The main conclusion from this is that hypoxia may not represent a clinical problem when small fraction doses are used over the range where the biphasic curve approaches the aerated slope. It follows from this, that any benefit from the use of hyperbaric oxygen is more likely to be found with larger fraction doses over the range where the biphasic curve deviates from the aerated slope. Here again a single value for OER cannot be applied over the range of fraction doses used in radiotherapy.

Having said that, it must be remembered that all the survival curves described in this chapter show the response to *single* doses. In a typical radiotherapy regime, such curves would only be applicable to the *first* treatment and a different situation may apply during subsequent treatments. Thus, the hypoxic fraction may rise or fall depending upon the rate of reoxygenation during the course of radiotherapy (Reoxygenation is discussed in Ch. 13) so that the efficacy of hyperbaric oxygen may also be variable.

Relative biological efficiency

An analogous situation exists when survival curves are compared after irradiation of cells with X-rays and fast neutrons. Figure 6.7 shows a pair of survival curves of HeLa cells irradiated in air with 300 kV X-rays and 14.7 MeV monoenergetic neutrons from a D-T generator. The curves are obviously different; the extrapolation numbers are 3.2 for X-rays and 1.6 for neutrons. This provides another example of variation in biological response over the range of doses used for fractionated radiotherapy. This time, however, the parameter which varies is the Relative Biological Efficiency (RBE) of fast neutrons compared with medium voltage X-rays.

Table 6.3 provides the analogy to Table 6.2, but for RBE instead of OER. While the OER value rose as dosage rose (and survival fell) the opposite is seen to occur with RBE. The data in Table 6.3 were obtained from the curves in Figure 6.7 where the X-ray survival curve is similar to that shown for aerated HeLa cells in Figures 6.1 and 6.5 (as might be expected).

The important conclusion to be drawn from these survival curves is that,

Table 6.3 Variation or RBE with survival level.

Survival level (per cent)	Neutron dose (rads)	X-ray dose (rads)	RBE
90	15	65	4.3
80	30	115	3.8
70	45	160	3.6
60	60	200	3.3
50	80	240	3.0
40	100	280	2.8
30	125	335	2.7
20	160	400	2.4
10	210	515	2.4
3	305	695	2.3

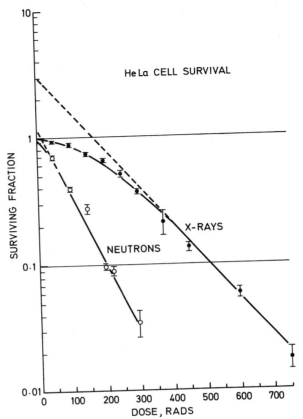

Fig. 6.7 Survival curves for HeLa cells irradiated with 300 kV X-rays or 14.7 MeV neutrons in air.

because of the relatively small shoulder on the neutron curve compared with that on the X-ray curve, the value of RBE changes considerably over just that range of dosage most commonly employed in fractionated radiotherapy. When fast neutron therapy is brought into routine use (because of the reduced OER) this variability will need to be remembered whenever a fractionated treatment is being prescribed, especially if it is proposed to choose a regime equivalent to some standard X-ray therapy regime. Each tissue may prove to have differences in the exact values for RBE, but the fact that RBE varies with different dose levels has been shown to be generally applicable. (Discussed in Ch. 15).

Cell survival through the cell cycle

Under the special conditions of a cell culture laboratory, it is possible to manipulate a cell population so that all the cells are passing through the phases of the cycle in synchrony (see Ch. 5). It is then found that, with the exception of mitosis, cells are most radiosensitive just towards the end of the G_1 phase. They then become progressively more resistant as they progress through the S phase but become more sensitive when they pass into the G_2 phase before the next

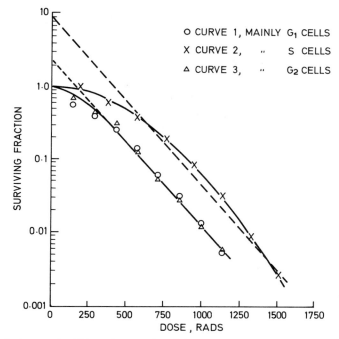

Fig. 6.8 Survival curves for Chinese hamster (lung) fibroblasts irradiated with 250 kV X-rays in air during the three phases of the cell cycle (from Sinclair, 1966).

mitosis. When this experiment is done under ideal conditions, the ratio of surviving fractions can amount to a factor of 40. If this large factor could even be approached under clinical conditions the therapeutic ratio might be very usefully increased.

It would be more realistic to look at the sort of survival curves shown in Figure 6.8 where the range of radiosensitivity is only 4-fold. There is some doubt whether the S phase curve is parallel to the G_1 and G_2 curves. Under clinical conditions this 4-fold ratio might still be useful and is one possible explanation for the synergistic effect of radiation combined with certain chemotherapeutic agents. Chemical synchronisation was discussed in Chapter 5 with agents like methotrexate as an example: cells are delayed in their progress through the cell cycle at just about the most sensitive point (end of G_1 — beginning of S).

It is important to stress that it is not the whole of this or that phase of the cell cycle that is radiosensitive or radioresistant, but rather that sensitivity varies throughout the phases of the whole cycle. It is quite common to hear the generalisation that G_1 cells are more sensitive and S cells more resistant. This is quite misleading since, as was said above, it is the particular point in the cycle when cells are completing G_1 and beginning S (the G_1 — S transition) that is the most radiosensitive. There are some cells with a long G_1 phase which are as radioresistant near the beginning of G_1 as they are near the end of S. This radioresistant portion of the cell cycle is absent, however, from cells with only a short G_1 phase, which progress quickly from mitosis to the G_1 — S transition point; both of which are radiosensitive.

Interphase death and lymphocytes

Most of this chapter has been devoted to *reproductive death* of irradiated cells, expressed in their inability to progress through enough cell divisions to form a colony in a petri dish or a spleen, or regenerate a crypt in the intestinal wall (to mention the examples quoted). For the majority of human cells, irradiated to the majority of clinical dose levels, this is the important manifestation of radiation damage. Loss of reproductive capacity is certainly the object of the exercise as far as tumour cells are concerned and it is the limiting factor with normal cells in the treated volume. If very high doses are delivered (e.g. thousands of rads within a few minutes) then radiation could cause the rapid cessation of cellular metabolism and then cellular disintegration. This is not a likely phenomenon in clinical practice. At lower dose levels, however, those cells which do not normally divide in adult life (like nerve and muscle) must necessarily die in interphase if they are to exhibit any radiation damage.

At such a clinical dose level, interphase death does occur in small lymphocytes and thymocytes. Because the usual tests for survival do not apply, it is difficult to find a relevant number for the radiosensitivity of these cells, but 50 per cent of a population of human lymphocytes 'look' dead within 24 hours of a dose of 100 rads. Such a dose delivered to all the other cell types discussed in this chapter would show no obvious effect other than mitotic delay until several days had elapsed and even then the surviving fraction would be around 80 per cent. When the short-term survival of lymphocytes was assayed after irradiation in organ cultures of whole lymph-nodes, however, dose-response curves showed a typical pattern with a D_0 value of 150 rads (and an OER of 2.7). Now that immunologists are recognising a variety of functional types of 'lymphocyte' these two very different estimates of radiosensitivity will have to be taken into account.

It is because the small lymphocyte does not normally go through a reproductive cycle in the peripheral blood that morphological indices of interphase death have been used to indicate radiation response. A special technique can be used, however, to stimulate at least one type of small lymphocyte (the 'T' cell, in the jargon of cell-mediated immunity) so that it transforms into a cell which goes through at least one mitotic cycle, including the period of DNA synthesis. Phytohaemagglutinin is used to stimulate the lymphocytes from their resting, or G_0 state. Thymidine labelling of the transformed cells can then provide an index of the degree of transformation, and this index has been found to be depressed if the G_0 cells are irradiated. With this technique a radiation dose-response curve has been obtained for small lymphocytes which is quantitative, although it still does not assay reproductive capacity in the usual sense. (The 'dose-response' curve for such lymphocytes had a D_0 value of 400 rads.)

Conclusion

Single dose survival curves provide an important example of the sort of information which radiotherapists may obtain from the radiobiology of single cells. Such curves have been shown to provide very little explanation for differences in the radiosensitivity of tumours — the curves illustrated here show a relatively narrow range in the parameters as listed in Table 6.1. The parameters for other survival curves are listed in Table 6.4 for comparison. These are for all

Table 6.4 Comprehensive list of survival curve parameters.

Cell Population	Assay	D_o	D_q	N	page
HeLa	*in vitro*	150	160	3.2	54
Chinese hamster (ovary)	*in vitro*	200	210	3.0	56
Chinese hamster (lung)	*small spheroid*	170	400	10	60
	large spheroid	170	900	100	60
Mouse leukaemia	*in vivo*	100	115	3.0	58
Mouse marrow	*in vivo*	100	100	2.5	91
Mouse marrow	*in vitro*	105	95	2.5	—
Human marrow	*in vitro*	137	0	1	—
Rat lymphocytes	*in vitro*	150	0	1	66
Human lymphocytes	*in vitro*	400	0	1	66
Mouse capillary endothelium	*in vitro*	200	160	2.3	87
Rat capillary endothelium	*in vivo*	170	340	7	128
Mouse skin	*in vivo*	135	350	—	84
Mouse melanoma	*in vitro*	133	190	4.2	166
Mouse small intestine	*in vivo*	130	450	—	59
Mouse stomach	*in vivo*	137	550	—	166
Mouse spermatogonia	*in vivo*	180	270	—	140
Mouse oocytes	*in vivo*	91	62	2	141
Rat thyroid	*in vivo*	405	400	2.8	138
Human neurones	*in vivo*	130	90	2	130
Mouse mammary carcinoma	*in vivo*	340	230	10	121
Rat rhabdomyosarcoma	*in vitro*	120	300	10	118

the cell populations discussed in this book. Other surveys in the literature force the additional conclusion that there is no systematic difference between normal and tumour cells with respect to these parameters of radiosensitivity. The search for improvements in the therapeutic ratio must lie elsewhere than in differences between the intrinsic radiosensitivity of cells; for example in the different growth kinetics of cell populations discussed in Chapter 10.

Since radiotherapy seldom involves single doses, these survival curves must be interpreted with care. They provide useful information when two or more situations are compared (e.g. the X-ray/Neutron comparison in Figure 6.7) and when two or more cell types are compared (e.g. Tables 6.1 and 6.4). It has been shown quite often enough that these curves can be generally applied to those clinical situations where they are relevant. The general principle of an initial *shoulder* followed by an *exponential* dose-response serves to illustrate the obvious truism that the larger the tumour cell mass the larger the radiation dose requirement. The biological application of these curves to fractionation schedules of radiotherapy is the subject of the next chapter.

REFERENCES

Bush, R. S. & Bruce, W. R. (1964) The radiation sensitivity of transplanted lymphoma cells as determined by the spleen colony method. *Radiation Research,* **21,** 612-621.

Hewitt, H. B. & Wilson, C. W. (1959) The effect of tissue oxygen tension on the radiosensitivity of leukaemia cells irradiated in situ in the livers of leukaemic mice. *British Journal of Cancer,* **13,** 675-684.

Paterson, R. (1963) *The treatment of malignant disease by radiotherapy.* London: Edward Arnold Ltd.

Puck, T. T. & Marcus, P. I. (1956) Action of X-rays on mammalian cells. *Journal of Experimental Medicine,* **103,** 653-666.

Sinclair, W. K. (1966) Radiation effects on mammalian cell populations in vitro in *Radiation Research.* Amsterdam: North-Holland Publishing Company. 607-631.

Till, J. E. & McCulloch, E. A. (1961) A direct measurement of the radiation sensitivity of normal mouse bone marrow cells. *Radiation Research,* **14,** 213-222.

Withers, H. R., Chu, A. M., Reid, B. O. & Hussey, D. H. (1975) Response of mouse jejunum to multifraction radiation. *International Journal of Radiaton Oncology,* **1,** 41-52.

FURTHER READING

Alper, T. (1975) ed. *Cell Survival after Low Doses of Radiation: theoretical and clinical implications.* London: John Wiley & Sons.

Altman, K. I., Gerber, G. B. & Okada, S. (1970) Radiation Biochemistry; *Cells,* **I,** by S. Okada. London: Academic Press.

Elkind, M. M. & Whitmore, G. F. (1965) *The Radiobiology of Cultured Mammalian Cells.* London: Gordon and Breach.

Friedman, M. (1974) ed. *The Biological and Clinical Basis of Radiosensitivity. Springfield, Ill.,* Charles C. Thomas.

7. Recovery from Radiation Damage

Except for the treatment of small skin lesions, radiotherapy is almost always given as a series of fractionated doses. Fraction doses of a few hundred rads are delivered to the volume, often at daily intervals for five days per week over a period of three to six weeks. For this reason the single dose survival curves described in the last chapter have only a limited application to the clinical situation, except as a basis for comparison of the radiosensitivity of different cell populations exposed to different forms of radiation under different environmental conditions. In this chapter the biological consequences of the fractionation of radiation dosage will be described. (The clinical consequences are discussed in Ch. 13.)

The three 'Rs' of education (reading, writing and arithmetic) have a radiobiological equivalent: *Recovery, Repopulation* and *Reoxygenation.* A fourth 'R' is often added: *Redistribution* in the cell cycle, to draw attention to the partial synchronisation of a cell population by irradiation (although this has a limited effect if several fraction doses are given). A fifth 'R' is the *Recruitment* of resting, G_o, cells into the growth fraction. Reoxygenation will be discussed in Chapter 10 (the response of tumours) although the effect of the oxygenation of a cell population will be shown to be of importance in the main topic of this chapter: *Recovery.*

During fractionated radiotherapy, both recovery and repopulation will tend to reduce the effectiveness of the total radiation dosage. Repopulation was discussed in the context of tumour cell kinetics in Chapter 5, but reference was made to radiation dose-dependent division delay in the previous Chapter 4. This may be one reason why repopulation is a less important factor in fractionation than recovery. Some experimental animal studies certainly suggest that recovery is more important. Pigs were irradiated either with one single dose or with five fractions in four days, or five fractions in twenty-eight days. The doses required to produce the same skin reaction are shown below in Table 7.1.

Table 7.1 The doses required to produce the same skin reaction

Fractionation	Total dose	Dose increment
1 Fraction	2000 rads	—
5 Fractions in 4 days	3600 rads	1600 rads
5 Fractions in 28 days	4200 rads	600 rads

These observations lead to the conclusion that even fractionation over the shortest period necessitates 1600 rads extra to produce the same effect. This is mainly attributable to recovery since repopulation would be minimal over that four day period; and even over the 28 day period the dose increment for the same number of fractions amounts to only another 600 rads. The recovery phenomenon was obviously more important than repopulation in that situation. Nevertheless, when large numbers of small fraction doses are given, so that recovery is quite small after each dose, then repopulation may then be comparable to, or even greater than, recovery. This would apply to comparisons of, say, 25 with 35 fractions given, say, over five or seven weeks; not to small numbers of large doses as in the above comparison of 5 fractions with a single dose.

Recovery from radiation damage is thus the first of the four 'Rs' of radiobiology. Cells may recover from *sub-lethal damage* and from *potentially lethal damage*. Both these phenomena influence the dose-response of mammalian cells to fractionated radiotherapy. Recovery is also a factor in that extreme form of fractionation, namely the protracted irradiation from radium and similar sources used for implantation and intracavitary treatment at a relatively low dose-rate.

A sparing effect of radiation when fractionated or given at reduced dose-rate has been known for many years and demonstrated in whole animal studies in mammalian cell culture systems *in vitro* and *in vivo*. In animal experiments, there appears to be a general similarity between the effects of fractionated doses and those resulting from a lowered dose-rate. At the single cell level, cultured Chinese hamster cells have been shown to be capable of repeated cycles of recovery from sublethal damage when the dose is fractionated. Figure 7.1 shows this in a schematic diagram which compares the surviving fractions of

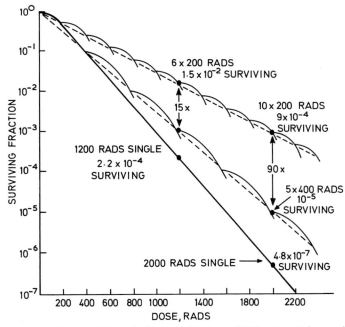

Fig. 7.1 Diagram of cell survival from single doses, fractions of 400 rads each dose and fractions of 200 rads each dose.

cells after either a single dose of 2000 rads (4.8×10^{-7} cells) or five fractions of 400 rads (10^{-5} cells) or ten fractions of 200 rads (9×10^{-4} cells). The cell survival curve is drawn in full for the single dose regime; only the 'shoulders' are drawn for the two fractionated regimes. These shoulders have exactly the same shape as that for the single dose curve. The point is that they must be 'reconstructed' after each fraction dose and this is the reason why larger numbers of fractions are less effective. Figure 7.1 shows that 10×200 rads is ninety times less effective than 5×400 rads; and 6×200 rads is fifteen times less effective than 3×400 rads. In each case, the same total dose is most effective if given in one single treatment.

The shoulder on cell survival curves

Since the existence of sublethal damage in an irradiated cell population is often equated to the presence of a shoulder in the cell survival curve, we must choose between a multi-target single hit or a single target multi-hit model.

The lethal effects of radiation may be explained in terms of a number of sensitive target structures within the cell being inactivated by single ionising events or, alternatively, a single target which requires to sustain multiple ionisation 'hits' before the cell loses its clonogenic capacity. This was discussed in Chapter 2 but the general conclusion is that although the multi-target model is often assumed by drawing an exponential curve after an initial shoulder, the cell survival data are often insufficiently precise to exclude the multi-hit model. Both models are compatible with a mechanism of *sublethal damage* when either the number of targets hit or the number of hits per target is below that required for lethal damage. The presence of a component of 'single hit', or irreparable, injury is shown by a definite slope on the survival curve at small doses, instead of zero slope. Few survival curves can be fitted to the above models without the addition of a proportion (at least 10 per cent) of 'irreparable' injury.

In addition to these mathematical models, the possible heterogeneity of a cell population may influence the size of the shoulder and may even result in the shoulder being lost altogether. It is therefore possible that sublethal damage may exist whether or not a shoulder is present in the survival curve. Even when the data can be fitted to a cell survival curve with a shoulder this does not provide direct evidence for the existence of sublethal damage in the system. While the presence of a shoulder on the single-dose survival curve may provide indirect evidence for the production of sublethal damage in some of the irradiated cells, only a fractionation experiment can demonstrate that these cells have the capacity to recover fully from this damage. Under the usual circumstances of acute irradiation of aerated cells, recovery from sublethal damage is so much the rule that it has been tempting to assume that the existence of a 'typical' single dose survival curve is not only evidence for the existence of a component of sublethal damage, but also that it can be assumed that the cells will recover from this damage.

The direct evidence for recovery from sub-lethal damage is provided by the sort of 2-dose experiment shown diagrammatically in Figure 7.2. A cell

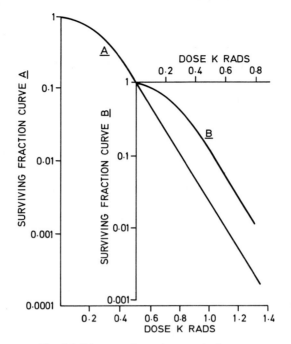

Fig. 7.2 Diagram of two dose survival curves.

population is shown to have a dose-response as in Curve A. Then Curve B shows that a similar dose-response will occur if the 10 per cent of cells that survive the first dose of about 500 rads are exposed to second doses. The shoulder is 'reconstructed' before the exponential slope of the second curve follows that of the first. A larger total dose is thus required to achieve the same biological effect if a time interval is left between the two fractions.

The extra dose required amounts to the size of the reconstructed shoulder. For a given biological effect (like reduction to a surviving fraction of 0.01) the single dose required will be D_1 while the total dosage required from two doses will be D_2. The dose increment is then $D_2 - D_1$. This value is usually the same as D_q, the quasi-threshold dose (discussed in the last chapter) which was one parameter for describing the size of the shoulder. This D_q might loosely be called the 'wasted' radiation before the more effective exponential dose-response becomes operative. Every fraction has this D_q and the many shoulders drawn in figure 7.1 illustrate the reduced effectiveness of fractionated radiotherapy attributable to recovery. If, however, each dose fraction is so small that it has not caused cell killing to reach as far as the end of the shoulder, then the recovered dose would obviously be less than $D_2 - D_1$.

D_q or $D_2 - D_1$ (which is the actual observation used for tissues *in vivo* discussed in Chapter 8) is the physical parameter of dose increment needed to produce the same biological effect. Alternatively the reduction in biological effect from giving the *same* total dose in two fractions may be measured. This is called the *Recovery Factor* and is illustrated in Figure 7.3 where the effect of

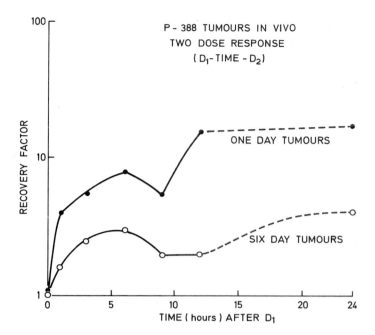

Fig. 7.3 Two dose response curves of ascites tumours (from Belli *et al*, 1967).

two equal doses $D_1 + D_2$ is shown. When both doses are given as one single exposure the survival level falls to 1 per cent. With an increasing interval of time between the doses the surviving fraction rises towards 10 per cent — the Recovery Factor is thus 10 and this is the biological consequence of fractionating a given dose. (The details of Fig. 7.3 are discussed later).

This recovery factor will often have a similar value to the extrapolation number (N) of the cell survival curve. N was the alternative parameter to D_q for describing the size of the shoulder of the survival curve. Just as D_q has loosely been called 'wasted' radiation, so N has sometimes been equated to the number of sensitive sites which must receive one hit before a cell is lethally damaged. This is a gross over-simplification of *target theory* for which there is very little direct evidence. It just happens that the Recovery Factor often coincides with the extrapolation number when these two parameters can be obtained for mammalian tissues or cell populations.

Again, the reservation must be added that the full Recovery Factor is only achieved if each dose is big enough to be beyond the shoulder region, as in Figure 7.2. In the absence of a more definite explanation for recovery from sub-lethal damage, this idea that N targets must be hit before the dose-response curve becomes exponential provides a working hypothesis. Studies with synchronised cell populations support this hypothesis since recovery is more evident later in the S phase of the cell cycle than earlier in the G_1 phase. There is obviously more of the important biological target, DNA, late in the S phase.

Table 7.2 Recovery capacity of mouse cells

Cell Type	D_q (rads)
Normal bone marrow	100
Leukaemic bone marrow	115
Mammary carcinoma	230
Osteosarcoma	280
Skin	350
Intestine	450
Stomach	550

All this implies that recovery from radiation damage, measured by increase in the biological parameter, survival (or decrease in the physical parameter, dose to produce the same biological effect) is synonymous with repair of damage to DNA. On the other hand there is, as yet, no direct evidence to equate repair of damaged DNA with Recovery.

Response of cells to densely ionising radiation

The cell survival curve for fast neutrons shown in Chapter 6 (Figure 6.7) had a much smaller shoulder than that for X-rays. From all that has been said, it might be expected that there would be less recovery with fast neutrons — if the amount of recovery is related to the size of the shoulder. This is indeed the case but the general statement is better worded in terms of the LET value of the radiation than the size of the shoulder. This is because there are many exceptions to the rule that the recovery factor in an irradiated cell population is equal to the extrapolation number of its survival curve. What can be stated, however, is that the higher the LET of the radiation, the smaller the shoulder of the survival curve and more particularly, the less the recovery from sub-lethal damage. The implications of this smaller shoulder were discussed in Chapter 6 in the context of RBE values, which will change significantly over the range of therapeutic doses.

The biophysical relationship between LET and RBE was discussed in Chapter 2 and one aspect of *target theory* was illustrated in Figure 2.4 there. The densely ionising radiation from alpha particles was shown to deliver many hits to the whole biological target (or cell) compared to the average of one hit or less from conventional sources of radiation. In terms of the over-simplified target theory discussed in the previous section, high LET radiation is so much more likely to hit *all* the sensitive sites in the cells that there is much less chance of any cells surviving with sub-lethal damage from which they may recover.

It might be said that there is a higher proportion of lethal damage, i.e. less sub-lethal damage, or that there is less recovery from sub-lethal damage. For clinical purposes this distinction is unimportant although the fundamental radiobiologist needs to investigate the phenomenon and clinical advantage may well accrue from any regimes which are shown to improve the amount of recovery of limiting normal tissues and/or reduce the recovery of malignant tissues. One example has already been mentioned in Chapter 4 *viz* the halogenated pyrimidine IUdR reduces the size of the shoulder on the survival curve and there is an associated reduction in the amount of recovery. This was an example of biochemical radiosensitisation and the use of more densely ionising radiation might be described as physical radiosensitisation. The same mechanism may be said to apply, namely a reduction in recovery from sub-lethal damage. All the same, sensitisation of malignant and normal cells to equal extents would give no therapeutic gain.

The radiotherapeutic consequence of this, is that less radiation is 'wasted' when high LET sources are used and this applies particularly over the lower dose range commonly used in fractionated radiotherapy (see Table 6.3). It is over that range where the 'shoulder effect' of conventional low LET sources is most evident so that the relative absence of shoulder with high LET sources is most apparent in fractionation regimes which use a larger number of smaller doses. With low LET radiation the upper curve sequence in Figure 7.1 applies and the total dosage must be considerably increased for the same biological effect. With high LET radiation this difference is very much less and fractionation formulae (see Ch. 13) have smaller terms as a result. This must make high LET radiotherapy much easier to use when different fractionation schedules come to be chosen (e.g. for specific effects on certain types of tumour).

Recovery from sub-lethal damage (SLD)

Returning to conventional radiotherapy sources which generate low LET radiation the response of mammalian cells *in vitro* to fractionated doses can be summarized as follows: An exponentially growing cell population contains cells at all stages in their growth cycle which differ in sensitivity to single doses of radiation (see Ch. 6). An acute first dose preferentially kills cells in the most sensitive stage. It follows therefore that those which survive are more resistant, and these become partially synchronized. They also carry sub-lethal damage. Intracellular recovery from sub-lethal damage leads to an increase in survival, but cell progression leads to a decrease following a second dose, since cells surviving the first dose must now progress to a more sensitive stage. These opposing effects lead, in terms of the two-dose response curve to an oscillating pattern (Fig. 7.3), a maximum followed by a minimum followed by further fine structure as the survivors pass towards the next division as a partially synchronized population. Except for the initial short interval between doses, the shape of two-dose recovery curves can be explained by the progression of a partially synchronized population through stages of varying radiosensitivity in the cycle. The initial sharp rise in survival is due to intracellular recovery from sub-lethal damage. However, it is not possible to determine whether recovery from sub-lethal damage is complete by the time of the first maximum, or has a time constant of 1 to 2 hr; in which case it would not become nearly complete until 4 to 8 hr after the first dose.

The oscillating patterns shown in Figure 7.3 showed that it took six hours for the first wave of recovery to reach a maximum in the mouse ascites tumours. These provide models of clinical situations because the tumour cell population is well oxygenated on the first day but poorly oxygenated by the sixth day. Quite obviously from Figure 7.3, there is more recovery in the oxygenated than in the hypoxic tumour. Recovery from sub-lethal damage in cells exposed to different conditions of oxygenation during and after irradiation has been the subject of much study and it can be concluded that cells which are severely hypoxic during and after irradiation suffer less recoverable damage than aerated ones. However, survival curves for cells irradiated under hypoxic conditions can rarely be drawn with extrapolation numbers significantly less than those for aerated ones particularly when statistical analysis is applied. The survival curves in Figure 6.1 showed that a full shoulder was maintained when the cells were irradiated under moderately hypoxic conditions. On the other hand the 'hypoxia' curve in Figure 6.5 had no shoulder; it was purely exponential and 2-dose experiments under those conditions of more severe hypoxia showed no recovery at all. It would appear therefore that not only do hypoxic cells possibly suffer less sub-lethal damage, but they are also less able to recover from the sub-lethal damage sustained. Both these processes appear to be affected by the degree of hypoxia during and particularly after irradiation, and therefore comparison is difficult unless the conditions are very precisely defined.

In summary, therefore, recovery from sub-lethal damage is probably a metabolic process since although it is relatively insensitive to small reductions in temperature and to low oxygen tension, (e.g. hypoxic cells at 340 ppm appear to be able to recover) it can be suppressed by some metabolic inhibitors, in particular those which are incorporated into DNA (IUdR, BUdR) or bind to the DNA (actinomycin D) (see Biochemical sensitisers in Ch.4). It occurs fairly rapidly with a time constant of about one hour and it involves reconstruction of the shoulder of the survival curve. A two-dose (or 'split-dose') experiment is required to demonstrate it but the amount of recovery is not dose-dependent. Less recovery is seen in hypoxic cells, provided they remain hypoxic during the interval between the two doses; the recovery then appears to occur more slowly. Less recovery is also found after high LET radiation.

Recovery from potentially lethal damage (PLD)

This other form of recovery has begun to attract attention because it may be more evident in cell populations which are not actively proliferating. These would be tissues with a negligible *Growth Fraction* (Ch. 5) which will include many tumours but also some normal tissues. Recovery from potentially lethal damage (PLD) has been demonstrated in cell cultures which have been allowed to grow into a densely crowded state. A series of radiation doses is delivered and the cells are then diluted and plated out for the usual assay of clonogenic capacity. If they are plated out straight away then a typical single dose-response curve is found. If, however, the cells are left in their crowded state for 6 or 12 hours without medium change before being plated out the survival curve is flatter (Fig. 7.4), i.e. survival is increased. In this case, the amount of recovery is dose-dependent, the higher the dose the more the recovery (in contrast to

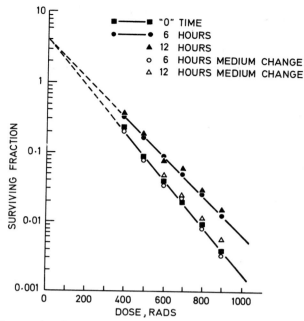

Fig. 7.4 Survival curves for plateau-phase cells with varying delays after irradiation (from Hahn and Little, 1973).

recovery from sub-lethal damage, SLD).

The phenomenon can easily be demonstrated in tissue cultures which can be manipulated to mimic the non-growth fraction situation found *in vivo*. Thus cells growing exponentially at 37°C do not recover from potentially lethal damage which is 'fixed' within a few minutes at that temperature. By contrast, if the same cells are cooled down to 20°C and held at that temperature, the damage is not fixed and recovery occurs. Thus the damage is only *potentially* lethal although it can be expected to be lethal in a proliferating tissue at 37°C. If a significant proportion of the irradiated tissue is not proliferating, however, there may be some of this form of recovery. Because it is dose dependent the phenomenon is less important over the usual range of fraction doses used in radiotherapy but *PLD* and *SLD* may be additive and some workers believe they are related phenomena. The time constant for recovery from *PLD* appears to be similar to that for *SLD*.

Whereas recovery from *SLD* is a fractionation phenomenon which occurs generally whenever two or more doses of radiation are delivered more than an hour apart, recovery from *PLD* is more difficult to demonstrate. If over-crowded cell cultures reflect a clinical situation this would be more likely to occur in the early stages of a course of radiotherapy when the tumour cell population has a lower growth fraction. Radiotherapeutic (and adjunctive) regimes which result in the majority of the cell population moving into proliferative activity will then be associated with a reduction in this form of Recovery, although recovery from SLD remains (as depicted in Fig. 7.1).

Finally, both forms of recovery occur during the first hours after a 'first' dose. The amount of SLD is measured by the response to a 'second' dose delivered at varying times after that first dose, whereas to measure the amount of PLD the important interval is that between the first dose and some stimulus which triggers more cell division. Both SLD and PLD are less evident after high LET radiation.

Recovery and dose-rate

Radiotherapists have grown accustomed to the use of sources of beam therapy which deliver tumour doses at a rate of about 100 rads/minute. At this 'conventional' dose-rate the fraction doses will be delivered over a period of just a few minutes and (assuming a recovery time constant of about one hour) recovery will be minimal during each actual treatment. Some radiation sources are capable of delivering higher dose-rates, (e.g. linear accelerators and the 'cathetron') and radiobiologists have investigated the possible advantages of increasing the dose-rate up to levels in excess of 10^{12} rads/minute. Because radiation is known to consume oxygen an 'ultra-high' dose-rate might deplete the oxygen tension of the cells in the irradiated volume so that, in effect, the whole volume would then be hypoxic and the therapeutic disadvantage of regions of poorly vascularised tumour cells would then be eradicated. The whole volume should show an equal radiation sensitivity.

This mechanism has indeed been shown to apply to bacterial systems in the laboratory but only at very high dose levels. With mammalian cells irradiated at a dose-rate as high as 10^{12} rads/minute single doses in excess of 2000 rads would be needed; a dose which is never likely to be employed clinically in a single session. At lower doses the mammalian cell studies at this ultra-high dose-rate showed no difference from the response at conventional dose-rates and for these reasons the clinical radiobiology of *increased* dose-rate does not seem to deserve further attention. When the dose-rate is *decreased* from a conventional level, the radiobiological consequences become much more significant, however.

All the survival curves discussed in Chapter 6 were obtained when cells had been given single doses of radiation at a conventional dose-rate, e.g. 100 rads/minute. The effects of such doses of 'acute' irradiation on single cells will reflect the results of individual doses of beam therapy in a typical fractionation regime. The effects of a radium implant or similar regime where the cells are irradiated continuously throughout a protracted period, may also be studied at the cellular level, using the test of survival of colony forming ability. Figure 7.5 shows two survival curves of Chinese hamster (ovary) cells; one curve for acute irradiation (100 rads/minute), the other for protracted irradiation (25 rads/hour).

It is clear that the cell survival curves become less steep and more nearly exponential as the radiation dose-rate is reduced. This is because cells recover from sub-lethal damage at a constant rate and this class of damage may fade to

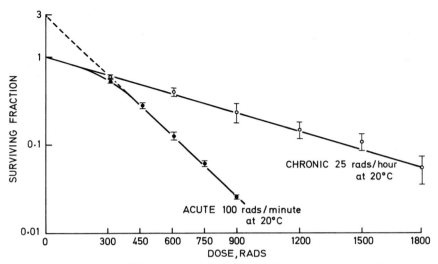

Fig. 7.5 Survival curves of Chinese hamster (ovary) cells after acute or protracted irradiation in air.

a negligible amount during protracted irradiation at a low dose-rate. The purely exponential shape of the curve for protracted irradiation in Figure 7.5 suggests that at that dose-rate (25 rads/hour) the cells do indeed shed sub-lethal damage as fast as they receive it. Thus the difference between the two survival curves is that the upper one indicates the dose-response with respect to lethal damage alone, while the lower one includes a component of sub-lethal damage from which the cells have not had time to recover before the remainder of the dose is delivered.

If this was the only difference in the biological effect of protracted radiation then there would be no particular advantage in using intra-cavitary and other arrangements of radium (or its substitutes) in radiotherapy, other than the very precise localisation of radiation dosage which can be achieved. There is some evidence for a radiobiological mechanism, however, which might explain the clinical success of traditional radium regimes. This suggests that protracted irradiation is more effective because the oxygen enhancement ratio, OER, is reduced at such very low dose-levels. Figure 7.6 shows two more survival curves for Chinese hamster cells after protracted irradiation, this time at 100 rads/hour. Both curves are exponential and the ratios of D_o thus indicate the OER which is only 1.4. This value is considerably less than the OER of 3.2 for these cells after acute irradiation (see Ch. 6).

There are difficulties in obtaining such data because of the long period of time (up to 30 hours) during which the cells are subjected to the special techniques, especially that for the hypoxic cells. One of the main problems in the laboratory is to control the amount of repopulation which may continue at this low dose-rate, although such cell proliferation is less evident in the hypoxic state. In the clinical situation it is even more difficult to measure the OER at low dose-rates but this radiobiological mechanism of a reduced OER might be expected to follow from the fact that recovery from sub-lethal damage pro-

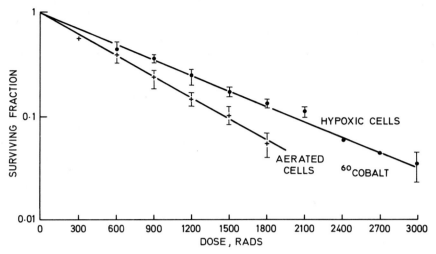

Fig. 7.6 Survival curves of Chinese hamster (ovary) cells after protracted irradiation in air or under hypoxic conditions.

ceeds faster in aerobic than hypoxic cells, so that the survival curves will tend to approach each other as in figure 7.6.

Proliferation of a cell population during protracted irradiation is even more likely to occur at very low dose-rates, below the clinical range used in radium implants. Figure 7.7 illustrates the result of exposing HeLa cells to a dose-rate of 2 rads/hour and then plating samples to measure the surviving fraction at 5

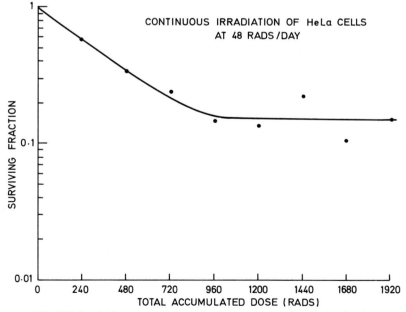

Fig. 7.7 Survival curve of HeLa cells after protracted irradiation in air.

day intervals. To start with the survival level falls, but after 20 days a plateau is reached where there is a 'steady state' between cell death and cell proliferation. Clearly both the processes of recovery and repopulation are evident at this dose-rate.

At even lower dose-rates the same phenomena will occur but the cell population will now show the minimum of perturbation from the normal level even when the population has been exposed to an accumulated dose of many hundreds of rads. This effect of very low dose-rates is mentioned because it applies to scattered and background irradiation, and the problem of radiation protection with which the Radiotherapist should be familiar. Until there is good evidence for a 'safe' lower limit of radiation dosage it must be assumed that *any* dose, delivered at any dose-rate, will be associated with cell damage; even though such damage may be difficult to demonstrate at the time. (The late effects of radiation are discussed in Ch. 12.)

Conclusion

Recovery from radiation damage is an important process of radiobiology which applies to the majority of present day regimes of radiotherapy where low LET radiation is employed and recovery will occur. The amount of this recovery seems to vary from one cell type to another. The parameter D_q is one index of this and (in the mouse) values for D_q extend from 100 rads for normal bone marrow cells to 550 rads for stomach epithelial cells (Table 7.1). The detailed shape of the shoulder region is even more important than the size of D_q when small fraction doses are considered; as is usual in radiotherapy. This variation in the amount of 'wasted' radiation may turn out to be as important in determining the therapeutic ratio between normal and malignant tissues as is the intrinsic radiosensitivity of their cell populations, as measured by the parameter *mean lethal dose, D_o.*

REFERENCES

Belli, J. A., Dicus, G. J. & Bonte, F. J. (1967) Radiation response of mammalian tumor cells. I. Repair of sub-lethal damage in vivo. *Journal of the National Cancer Institute*, **38**, 673-682.
Hahn, G. M. & Little, J. B. (1973) Plateau-phase cultures of mammalian cells: an in vitro model for human cancer. *Current Topics in Radiation Research*, **8**, 39-83.

FURTHER READING

Elkind, M. M. & Sinclair, W. K. (1965) Recovery in X-irradiated mammalian cells. *Current Topics in Radiation Research*, **1**, 165-220.

8. Early Response of Normal Tissues

Radiotherapists need to understand the response of those normal tissues which are likely to limit the dosage which is primarily intended for the tumour. The term 'radical' radiotherapy implies an attempt to cure a tumour by delivering the maximum dosage which the irradiated *normal* tissue can tolerate. The maximum dose which may safely be given to a particular type or volume of tissue is known as the *tolerance dose*. The normal tissues to be discussed in this chapter are therefore those which commonly limit radiotherapy in this respect. These 'limiting normal tissues' include *skin, gastro-intestinal tract, bone marrow* and *blood vessels*. Blood vessels may well be the most important of all, since damage to the vascular supply will eventually lead to damage to other tissues, discussed in Chapter 11 as Late Effects.

The *radioresponsiveness* of different cell populations was discussed in Chapter 3 where the kinetic parameters of the limiting normal tissues were listed in Table 8.1. Restating more precisely the old Law of Bergonié and Tribondeau, the cell populations with the higher rate of cell division showed the earlier response to radiation. Skin, intestine and bone marrow were seen to be more responsive, vascular and connective tissue were intermediate and the CNS was slowest to respond. In each case the 'biological target' is a cell. Any cell has only a statistical probability of being damaged by the ionising radiation and a proportion of cells will remain unscathed. Whether or not the damage becomes manifest as a significant depletion in the cell population depends to a large extent on the detailed characteristics of the renewal system of which that cell is a part. These characteristics will now be described for the limiting normal tissues, together with such quantitative information as is known for their radioresponsiveness.

Skin

The histopathological response of skin to irradiation was illustrated in Chapter 3 by three photomicrographs (Fig. 3.2) of skin sections. Skin is a composite organ and consists of an outer epidermal layer, a cutaneous and subcutaneous connective tissue layer, and various accessory organs, such as hairs, nails, exocrine glands, and sensory receptors. All of these structures participate to varying degrees in the reaction of skin to radiation exposure.

The epidermal layer represents a cell renewal system akin to the intestinal epithelium, but cell replacement and transit times are slower. Only about 2 per cent of all cells are renewed daily in the epidermis as contrasted with 50 per cent in the intestine. The epidermal cells divide in the germinal stratum, differentiate by keratinization, and are finally shed from the surface of the skin.

The length of the cell cycle and the transit time depend largely on the anatom-ical site, age, species, and functional state of the skin. Thus, transit time in the epidermis varies from about 14 to 17 days in man and 8 to 12 days in mice and rats. The cell cycle times are about 7 and 5 days respectively, and the transit times are related simply to cell cycle times because the epidermal layer is nor-mally 2 to 3 cells thick except on plantar and palmar regions.

One to two days after exposure to 800 rads or more, the skin may redden temporarily. During this early erythema the blood vessels are congested and oedema occurs in the subcutaneous layer, whereas the epidermis appears normal except for mitotic arrest. This initial erythema increases during the first week but fades to a minimum at the tenth day. The main erythematous reaction then becomes maximal on the fifteenth day and lasts 20-30 days after exposure. Figure 8.1 shows this on the basis of skin reflectance which provides a measure of the change in skin colour. The main erythema reaction involves not only the epidermis, but also the underlying strata of skin, especially the blood vessels. When the basal layers of the epidermis regenerate, the erythema disappears, but may reappear in a wave-like manner and lead to dry or moist desquama-tion.

Once desquamation occurs erythema measurements, like those in Figure 8.1, become meaningless and a quantitative measure of skin reaction can only be obtained by the use of an arbitrary scale extending from mild erythema to moist desquamation of the entire irradiated area. Trained observers can make useful comparisons using such a scale and the figures quoted at the beginning of Chapter 7 (Table 7.1) were based upon such comparisons. These are neces-

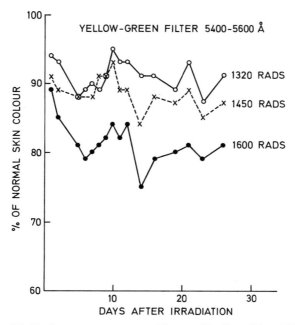

Fig. 8.1 Erythema measurements of human skin (from Nias, 1963).

sarily confined to the superficial effects upon irradiated skin. A biopsy will provide more information on effects throughout the full skin thickness but such biopsies cannot usually be repeated over a period of time (and obviously not from the same site) so that serial observations are not possible clinically.

At one point of time the effects of both a *tolerance dose* and an overdosage of radiation where shown in terms of the histological appearance in Figure 3.2. This was the response of the whole complex tissue which is skin; with all the elements shown there in Figure 3.2. It is difficult to quantitate such histopathological changes without selecting certain of the constituent cell populations for individual study. Such selection is necessarily artificial and may even be misleading if the relative importance of the particular cells to the integrity of the whole tissue is not understood. Despite this, two cell populations may be worth considering: the epidermal and the capillary endothelial. (The endothelial cell population has more general application and will be considered after the end of this section on skin.)

Figure 8.2 shows a dose-response curve for epidermal stem cells. This was obtained on the basis of the number of regenerating nodules formed in the

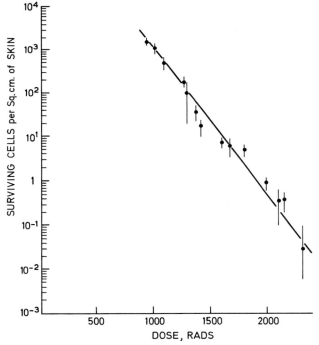

Fig. 8.2 Dose Response curve for mouse epidermal cells (from Withers, 1967).

denuded skin of mice after varying doses of irradiation. A special technique is used whereby increasing areas of skin are shielded from a lethal dose delivered first to a surrounding area to ensure that no stem cells can migrate into the area to be tested. Since only a limited number of nodules can be counted within a given area of skin, there is a lower limit of dose below which this method cannot be used to resolve the dose response. The curve in Figure 8.2 is limited to the exponential portion of what is presumed would be a 'shouldered' dose-response curve if the lowest dose-range could be resolved. This presumption is justified by the results of split-dose experiments which show that recovery occurs and that the D_q amounts to 350 rads (on the basis of a $D_2 - D_1$ calculation, see Chapter 7). This would be the extent of the shoulder of a complete curve.

The radiosensitivity of this mouse epidermal stem cell population can thus be shown to be typical of many other cell populations (Ch. 6) when measured *in vivo* in this way. Under aerated conditions the D_o value is 135 rads; under hypoxic conditions the D_o is 350 rads. The oxygen enhancement ratio is thus 2.6 which is also typical of other cell populations and this observation suggests that, in the mouse at least, skin is normally well oxygenated.

A further piece of evidence that epidermal stem cells show a 'typical' radiation response *in vivo* is shown in Figure 8.3. Here again the regenerating skin nodules have been counted after various doses of irradiation but this time the response to X-rays is compared with that to fast neutrons. The dose-response

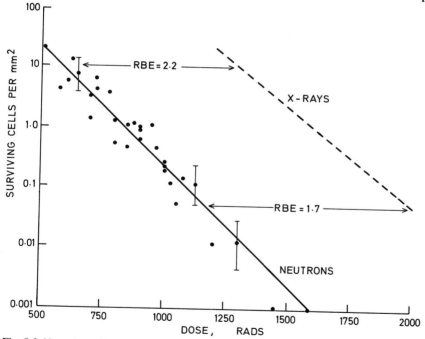

Fig. 8.3 Neutron and X-ray dose response curves for mouse epidermal cells (from Denekamp, Emery and Field, 1971).

curves have D_o values of 109 rads for neutrons and 125 rads for X-rays. The RBE values thus fall with higher doses; a trend which was shown for cultured cells in Chapter 6 (Figure 6.7 and Table 6.3) and will be shown to be generally applicable to all tissues irradiated with fast neutrons (Ch. 15).

With the increasing use of supervoltage radiotherapy the superficial skin reaction has become less important and, in this respect, skin may no longer be considered a 'limiting normal tissue'. The deeper layers of skin continue to receive maximal entrance doses, however, and these dermal layers will be involved in *wound healing* when pre-operative radiotherapy is used. It is difficult to separate the response of connective tissue from that of the parenchymal cells which are of prime interest. A functional test can be used, however, such as that shown in Figure 8.4 which shows the effect of irradiation

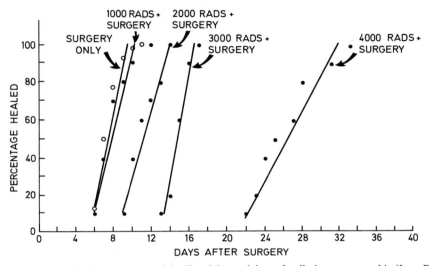

Fig. 8.4 Relation between wound-healing delay and dose of radiation to mouse skin (from Powers and Palmer, 1968).

upon wound healing in the mouse. A single dose of 1000 rads did not delay healing at all. With higher doses there was a delay which was dose dependent but all the wounds healed eventually. Clinical experience suggests that human skin responds in a similar way. The increasing interest in this particular regime of combination cancer therapy, namely *pre-operative radiotherapy* before radical surgery, requires a knowledge of the effects of radiation upon connective tissue in the operative site.

Blood vessels

In the section on vascular tissue in Chapter 3 it was stated that the larger vessels are apparently radioresistant, in contrast to the capillaries and small arteries which become occluded after moderate doses of irradiation. It will be damage to these smaller vessels which will delay wound healing (Fig. 8.4), and cause late damage to the skin (Fig. 3.2(c) and other tissues such as the kidney and the central nervous system. At the cellular level, the endothelial lining of large vessels is probably as sensitive as that in capillaries but, because of the large diameter of those vessels, even if endothelial proliferation and swelling or blood clotting does occur, the vessels will not be occluded. The radiosensitivity of capillary endothelium has been determined in the rat by a method in which the capillaries were induced to proliferate in a thin sheet of subcutaneous tissue. A value of 170 rads was found for the D_o of the survival curve for which the extrapolation number was 7.

Capillary endothelial cells from mouse kidney have also been studied, in a comparison using in vitro tissue culture techniques. 'Typical' cell survival curves were found (Figure 8.5) having values for D_o of 200 rads in air, 530 rads

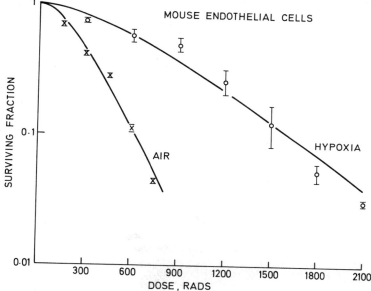

Fig. 8.5 Survival curves for mouse endothelial cells irradiated with 300 kV X-rays in air or under hypoxic conditions.

under hypoxic conditions, the same extrapolation number 2.3 and thus an oxygen enhancement ratio of 2.65. If these results are relevant to human blood vessels they show that vascular endothelial cells are just as radiosensitive as other cells.

In Table 3.1 of Chapter 3 vascular endothelium was shown to have rather slow tissue turnover times. Damage develops more slowly. The turnover time for endothelial cells in the capillaries of a mouse tumour is 50 hours (as compared to 22 hours for the tumour cells). Eventually, however, tissues and organs with the same order of kinetic parameters (liver, thyroid, connective tissue) and in the lower order of kinetics (e.g. CNS) will show radiation damage as an indirect result of vascular damage and this will be discussed in Chapter 11 (Late Effects).

Gastro-intestinal tract

In the whole digestive tract (the mouth, pharynx, oesophagus, stomach, small intestine, large intestine, and rectum) the small intestine is the most important site of radiation injury. All three regions, the duodenum, jejunum, and ileum are lined (Fig. 8.6) by a columnar epithelium consisting of mucus and columnar or 'chief' cells. The crypts which contain the generative cells for epithelial replacement are found in the mucosa at the bases of the villi. The cells of the crypts and of the related villi can be considered parts of a cell renewal system, which is in a state of kinetic equilibrium. Cell renewal occurs in the

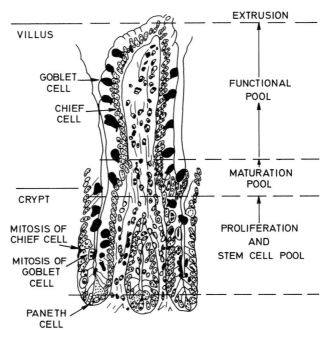

Fig. 8.6 Diagram of intestinal epithelial pools (from Bond, Fliedner and Archambeau, 1956).

mitotic areas in the crypts. From there, newly formed cells migrate out and move from the base of the villi to the top, designated as the extrusion zone. The predominant cell in this renewal system is the columnar cell and scattered among them are goblet cells. At the bottom of the crypts are the Paneth cells which show few alterations following irradiation.

The small intestine is the most radiosensitive portion of the digestive tract. The duodenum (portion nearest the stomach) is the most sensitive segment and the epithelium in the crypts is more sensitive than that of the surface. Thirty minutes after a moderate exposure, about half of the crypt cells of the duodenum show nuclear fragmentation, swelling, or other evidence of cellular disintegration. Debris accumulates in the lumen of the crypts. Within the first few hours most mitotic activity ceases. The cells which do undergo division have bridges or acentric fragments. Within a day, the epithelial surface contains only a thin layer of cells and the villi are shortened due to this reduction in the number of epithelial cells (see Fig. 9.9 in Ch. 9).

The consequences of a large dose of radiation to the intestines were described in the classical paper *The Nature of Intestinal Death* by Quastler (1956):

A mammal with a leaky intestinal barrier can survive. If a large segment of the intestinal barrier is missing for some time, however, then death is bound to occur. The problem of the immediate cause of death in acute intestinal radiation death boils down to the question of which one (or ones) of a number of likely mechanisms kills the animal before the others become effective. The timing is important; animals will succumb only if the recuperative activity in the crypts does not become effective before denudation is complete. If it were possible to tide the animal over the fastest dangerous reaction, then it might be possible to avert acute intestinal radiation death altogether.

. . . The loss of the intestinal barrier has three likely consequences: intestinal bacteria and their toxins can enter the submucosa and invade the bloodstream and the peritoneal cavity; proteolytic enzymes in the intestinal lumen can digest submucous tissue and enter into the peritoneal cavity; in the other direction, water and electrolytes will be lost into the intestinal lumen.

That description was applied to the mouse and much of our radiobiological information depends upon the study of that species, for both limiting normal tissues and tumours. This information is much more accurate than that obtainable in human studies which are necessarily limited for ethical reasons. Intestinal death in man will be discussed in the next chapter but the radiobiological principles which lead to it are similar to those described here for the mouse. Only the time scale is different, with the turnover time for human cell populations tending to be longer by a factor of about 2. The time of death is likely to be earlier in the mouse, but the mechanism of death is similar.

Recovery of the small intestine is rapid. There is evidence of mitotic activity 1 day after exposure but many of the new cells are abnormal. By the third day, the rate of mitosis in the crypts is greater than normal, although the villi are covered by only a thin layer of stretched cells. By the end of a week, new cells have covered the villi and the intestinal epithelium appears about normal. Removal of cellular debris is slow. Regeneration commonly occurs in the presence of degenerative changes. The initial damage is more severe following large doses and recovery of the epithelium may not then occur. The villi appear to be shortened and only partially covered by degenerating cells. Death of the animal usually occurs between the third and fifth day in the absence of regeneration of the epithelium of the small intestine.

In the past a number of radiobiological studies employed intestinal death as the end-point when comparing effects like, for example, X-rays and neutrons, or radiation at high and low dose-rates. Nowadays, such questions can be examined using the response of intestinal stem cells which was described in Chapter 6. A survival curve for mouse jejunum (Fig. 6.4) was shown with an apparently very large shoulder derived from fractionation studies. The final exponential portion of that curve is 'real', in the sense that the D_0 value of 130 rads is based upon the number of regenerating crypts of Lieberkuhn which can be counted around the circumference of a section of gut, when fixed and stained 3 to 4 days after single doses of radiation over the dose range above

1100 rads. The survival curves for epidermal stem cells (Fig. 8.2 and 8.3) were also restricted to a middle range of doses. In both cases the number of regenerating stem cells becomes too high to be counted when lower doses are administered. Estimation of the real size of the shoulder was described in Chapter 7 — split-dose experiments show a D_q value of 450 rads for intestine after high doses (and 550 rads for stomach where the D_o value is in the usual range; 137 rads). This is larger than the D_q of 350 rads for skin but since the D_o values for the skin and intestinal cell populations are very similar, 135 rads and 150 rads respectively, the more acute radiation response of intestine must be attributed to differences in the kinetics of the cell populations (the cycle time for intestinal stem cells is 12 hours; for skin it is 5 days).

Bone marrow

The bone marrow varies in structure and cellular compositon with age and with anatomical location. In the young individual it is distributed through all bones and acts as one single organ system. In the rodent, active marrow persists in the adult, particularly in the long bones. In man and larger animals, active marrow is confined to the flat bones (sternum, ribs, iliac crest) and the epiphyses of long bones.

It is known that the haematopoietic bone marrow is composed principally of three cell renewal systems: erythropoietic, myelopoietic, and thrombopoietic, but the morphological identity of the stem cell (or cells) of the bone marrow and even the anatomical boundaries of its normal site of origin and its location are not known. The stem cell (or cells) is known to exist however, and its proliferative capacity can be characterized by indirect physiological means. The mature granulocyte and lymphocyte may spend a sizable fraction of their life span outside of the blood vessels. There is good evidence to consider the division of this proliferative system into 'pools' and 'compartments' (Fig. 8.7)

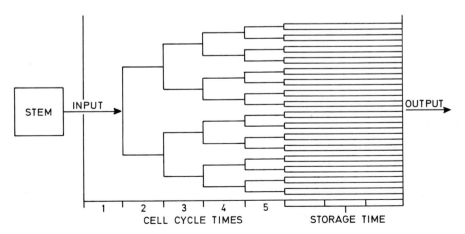

Fig. 8.7 Compartments of a cell proliferation system (from Lajtha 1965).

to characterize the normal and abnormal states and to allow systematic kinetic studies for the establishment of the key parameters of the system. But such division into *compartments* can only be arbitrary and will depend to some extent on the technique and the judgment of the observer. Survival of bone marrow stem cells can be measured by a variety of methods e.g. the spleen colony method (Ch. 6) and erythropoietin response. When a list is made of values of D_0 and N obtained by five different methods, D_0 varies between 60 and 105 rads, while N lies between 1.2 and 2.7. These values of D_0 overlap the range for cells cultured *in vitro*, although they fall at the lower end of the range.

The consequence of radiation depletion of the bone marrow stem-cell population is eventually seen in the form of a depression in the peripheral blood count. Figure 8.8 shows the peripheral blood counts of rats after 500 rads of whole body irradiation (see Fig. 9.7 in Ch. 9 for the polymorph count of monkeys after 820 rads whole body irradiation). Time elapses after the bone marrow is irradiated before the maximum depression occurs followed by recovery to a normal level. This time interval depends upon the kinetics of the various types of stem cell population. Each of the marrow populations have their own kinetics and the time scale of the perturbation in the peripheral blood count varies accordingly.

The earlier 'compartments' (Fig. 8.7) of these blood cell populations will

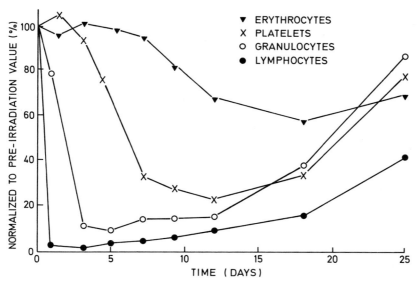

Fig. 8.8 Peripheral blood counts of rats following 500 rads whole body radiation.

involve proliferation in the bone marrow and the peripheral blood counts show only the 'output' compartment. The total period of time from input to output starts the moment a stem cell is stimulated to begin proliferation along one of the three possible pathways of differentiation: erythroid, myeloid, and platelet, (e.g. the myeloid pathway ending up in a mature polymorphonuclear leucocyte). The stem cells initiate a variable number of cell divisions during which differentiation and maturation gradually supervene before the cells appear in the peripheral blood in their recognisable form. Figure 8.9 compares the time scale of this maturation of the three cell populations and their subsequent life span in the peripheral blood, for rat, rabbit, dog and man. It is notewothy that red cells have a relatively very long life span in the peripheral blood for all species; 110 days for man. The other two cell types have both a shorter total life span (from start to finish) and a shorter period in the peripheral blood. It is the total time however, from start to finish, which determines the responses shown in figure 8.8. This will be further discussed in Chapter 9. Values for the

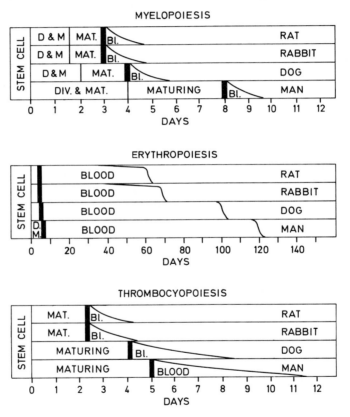

Fig. 8.9 Diagram of the time parameters for haemopoiesis in the rat, rabbit, dog and man (from Bond, Fliedner and Archambeau, 1965).

'turnover' and 'transit' times of these bone marrow cell types were contrasted to their life-span (and that of intestinal epithelial cells) in Chapter 5.

The peripheral blood count which shows the earliest depression of all is that for lymphocytes. This cell population is receiving increasing attention because recent advances in immunology are centred upon the phenomenon of cell-mediated immunity and the immune system must now be included in the category of limiting normal populations.

Cell-mediated immune system

Tumour immunology is one of the fastest developing subjects in biological science just now so that any detailed description may turn out to be either misleading or even wrong within a very short time. It may be helpful, however, to give a general description of those components of the immune system which are known to exist and can be monitored, even though their functional relationships are not yet completely understood. The immunological surveillance mechanism involves both lymphoid cells and antibodies which react with antigens. Antigens may or may not be specific to a particular tumour or its parent organ. For an immune response to be mounted, that antigen must first be recognised (as foreign protein) and this is believed to be the function of T-lymphocytes (the lymphocyte populations are discussed below). Following this recognition, an antibody, specific to the antigen, will be synthesised by B-lymphocytes in the form of immunoglobulin which can then circulate in the blood stream and react with the antigen. Really 'foreign' proteins are quickly rejected by a patient who is fully immunocompetent. Tumour cells are less effective in this respect, either because they are less antigenic (i.e. less 'foreign') or because the antibody response is blocked in some way. Any agent (like radiation) which damages any of these components may depress the immunological surveillance mechanism. Radiotherapy is thus an immunosuppressive agent and is regularly used as such for tissue transplantation.

Recent work on the organisation of the lymphoid system has shown that it includes 2 major cell populations called T and B lymphocytes, which differ in origin, in distribution within the peripheral lymphoid organs and in function during the immune response. Their interrelation and function are illustrated in Figure 8.10. This diagram shows the circulation of cells from (1) the bone marrow via (2) the thymus to either (3) Peyer's patches or (4) lymph nodes or (5) the spleen, and then by lymphatic drainage back into the blood to be recirculated. Both cell populations are believed to have a common origin in a multipotent haemopoietic stem cell, which may or may not be processed during differentiation through the thymus. The population that derives from the thymus is currently known as the T-cell population and constitutes the majority of the recirculatory pool of lymphocytes in areas generally designated as 'thymus-dependent'. The T cell population is of primary importance in the *induction* of cell-mediated immunity leading subsequently to antibody production by B cells. This is a process that requires co-operation between the two major cell populations and other classes of lymphocyte (e.g. 'killer' cells).

The bone-marrow derived population that does not require the presence of the thymus for differentiation, is currently known as the B or thymus-independent lymphoid population. These are the antibody-producing cells which are also capable of recirculating from blood to lymph, but to a lesser extent than the T cell population. As far as the life-span of these two lymphocyte populations is concerned it is possible to generalise that most long-lived lymphocytes are thymus-derived and most short-lived cells are of marrow

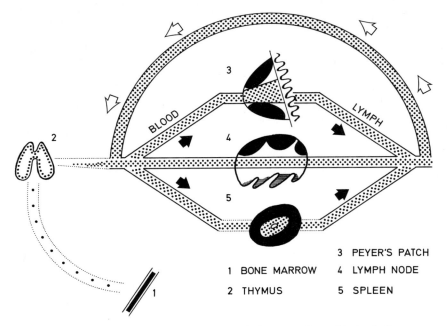

3 PEYER'S PATCH

1 BONE MARROW   4 LYMPH NODE

2 THYMUS   5 SPLEEN

Fig. 8.10 Diagram of cell mediated immunological system (from Maria de Sousa, personal communication).

origin; but there are exceptions to this rule. The short-lived cells acquire label within 4 to 5 days but many of the long-lived cells remain unlabelled even after multiple injections of radioisotope over weeks or months.

The continuous traffic of lymphocytes from blood to tissues and to lymph means that cells with the capacity for mounting a specific immune response may be present in any lymph node, in the spleen, in blood or in lymph. However, the lymph nodes draining the site of antigenic stimulus (e.g. the tumour) play an important part in initiating and maintaining the immune response, via their role in trapping and processing antigen. The spleen probably acts as the antigen trap for the bloodstream; the Kupfer cells of liver trap antigen from the portal system, but are unusual in that they process antigens in such a way as to reduce their immunogenicity. It is the very nature of cancer, however, that most tumour cells can only mount a weak immune response to start with.

The response of the immune system to radiation can be considered both in terms of a reduction in the populations of lymphoid cells which perform the function of cell-mediated immunity, and also in terms of a depression of the humoral immunity from antibody production by such cells. Following whole-body irradiation the B-lymphocytes are much more radio-responsive than the T-cells. The evidence for this is both morphological and functional; primary nodules of lymphoid cells will reappear within 7 days, but restoration of the depleted thymus-dependent compartment may still take 4 to 5 weeks. Both active antibody synthesis and circulating antibody are radioresistant, however. Antibody-synthesising cells (primarily the plasma cells), continue to produce antibody for several days after irradiation. The kinetics and radiation dose-response of bone marrow stem cell populations were considered earlier in this chapter.

Conclusion

The normal tissues whose early response limits the dosage of radiotherapy include skin, blood vascular endothelium, gastro-intestinal tract and bone marrow. In each case a cellular response can be identified with a typical dose-response curve. Much of the information derives from mouse and other experimental animal studies but the human response will vary only in time, because of kinetic differences between the relevant cell populations. Perhaps the most important cell population of all is that of capillary endothelium. Damage to capillaries and small arteries leads to damage to the tissues supplied by those blood vessels. Indirectly, therefore, the limitation of radiotherapy dosage depends upon the radiation response of capillary endothelial cells. The later consequences of this will be discussed in Chapter 11.

REFERENCES

Bond, V. P., Fliedner, T. M. & Archambeau, J. O. (1965) *Mammalian Radiation Lethality.* New York: Academic Press.
Denekamp, J., Emery, E. W. & Field, S. B. (1971) Response of mouse epidermal cells to single and divided doses of fast neutrons. *Radiation Research,* **45,** 80-84.
Lajtha, L. G. (1965) Response of bone marrow stem cells to ionizing radiation. *Current Topics in Radiation Research,* **1,** 139-164.
Nias, A. H. W. (1963) Some comparisons of fractionation effects by erythema measurements on human skin. *British Journal of Radiology,* **36,** 183-187.
Powers, W. E. & Palmer, L. A. (1968) Biologic basis of preoperative radiation treatment. *American Journal of Roentgenology,* **102,** 176-192.
Quastler, H. (1956) The nature of intestinal radiation death. *Radiation Research,* **4,** 303-320.
Withers, H. R. (1967) The dose-survival relationship for irradiation of epithelial cells of mouse skin. *British Journal of Radiology,* **40,** 187-194.

9. Acute Radiation Syndromes

The energy in a lethal dose of total body radiation is only sufficient to boil a teaspoonful of water. This small amount of energy is nevertheless sufficient, when delivered in the form of ionizing radiation, to have profound effects upon cellular systems in the body. The effects will depend upon the proportion of the body exposed to radiation and the dose delivered. A single acute dose of 2000 rads delivered to an area of skin 1 square centimetre will lead to localized moist desquamation but no systemic effects will be observed. Half this dose delivered to the whole mammalian body will lead to death within two weeks, after a prostating illness. Between these extremes, partial-body irradiation will produce different results (dose for dose) depending upon the particular tissues exposed. The initial effects produced by radiation may lead to clinical effects expressed promptly, or only months, or years after irradiation; depending, not only on the nature and extent of the initial radiation injury, but also on secondary factors, such as hormonal influences, or subsequent exposure to other noxious agents.

In this chapter, the somatic effects of acute radiation will be considered; especially the early effects, both of whole-body and of partial-body irradiation. These early effects will be related to the cellular damage which is inflicted in various parts of the body and it will be shown that the lethal and other profound effects of acute irradiation can be attributed to damage to specific cell populations. These populations are found in different parts of the body and they will be discussed first of all as part of the whole-body irradiation syndrome and then as part of different partial-body syndromes. Although the main emphasis should be upon effects in man, there is not a great deal of information which is accurate and quantitative enough about the effects of total-body irradiation in man. Such evidence as exists will be quoted but much of the discussion of the precise cellular relationship of the acute radiation effects must be extrapolated from the more detailed studies possible with other animals (see Ch. 8).

Although there has been very little deliberate total-body exposure of man, some evidence is available from a group of patients observed for 42 days after single doses of gamma radiation in the range 30 to 300 rads (Lushbaugh *et al.,* 1967). There have also been various industrial radiation accidents and there is the evidence from Hiroshima and Nagasaki, although all the accidental evidence is confused somewhat by difficulties of dosimetry in the individuals exposed (UNSCEAR Report, 1962). The best estimate of the median lethal dose (LD50) for man is 300-400 rads short-term total-body irradiation; the actual value depends on the quality and distribution of the radiation (The significance of the expression LD50 will be discussed on p. 101). This does not mean that man can tolerate this amount of radiation, since all individuals exposed to this level would have serious symptoms, and 50 per cent would die.

It must be stressed that the results of exposure to 200 rads short-term total-body irradiation may sometimes cause death.

Dose/volume relationships. Death from acute radiation results from damage to three functional bodily systems: the Central Nervous System, the Gastrointestinal System and the Haemopoietic System. The size of radiation dose and the proportion of each or all of the three systems exposed, detemines whether death occurs and the mode of death. A detailed discussion of the three syndromes will be given on p. 99. Other organs and systems are always involved simultaneously with those whose damage is, or may-be conducive to death. But not all observed changes are morphological; functional effects also occur, e.g. modification of conditioned reflexes in animals given local doses to the head as low as 5 rad. Other organs reveal damage at a later date and this comes under the heading of 'Late Effects'. These may eventually be lethal but are considered in Chapter 11.

As the volume of the body exposed to such acute doses of radiation becomes smaller, the problem comes more and more into the context of conventional radiotherapy where considerable knowledge is available of the effects of acute dosage on different regions of the body. A standard textbook of radiotherapy can be consulted for further information in this range (e.g. Paterson, 1963). Such a textbook would discuss the so-called tolerance dose; that is to say, the dose of radiation which a particular volume of the body may tolerate with apparently normal recovery. This must be considered both in relation to a simple mathematical volume and also to the particular region being irradiated. Some tissues, such as the kidney, have a more limited tolerance. Other tissues, such as skin, may tolerate more, but the rules of total volume of tissue exposed to the radiation are fairly general. This chapter is not concerned with the comparatively small volumes exposed to radiotherapy. It is total-body irradiation which will be discussed in detail below, and also such large partial-body radiation which is relevent to the total body effects.

The Total-body response in man

The total-body response in man to radiation includes (1) radiation sickness beginning during or very soon after acute exposure, (2) degeneration and repair of proliferative tissues, (3) local and generalised toxaemia, (4) changes in homeostasis, (5) deterioration in physical and mental fitness, (6) death. The various syndromes and their time scale are shown diagramatically in Figure 9.1.

This shows that manifestations of all the syndromes overlap to a certain extent, depending upon the level of dose delivered to the whole body. Prodromal nausea and vomiting will be found even after doses that do not prove lethal but will last longer after higher doses. Any CNS symptoms will also occur in the initial period after irradiation but these indicate a very high and lethal dose level. Gastrointestinal symptoms will usually indicate a fatal outcome also, but the time-scale is dose-dependent, as is the time of onset of haematological changes. These are not necessarily fatal over the lower dose range. The time scale extends from hours and days to a number of weeks. The longer an irradiated individual survives after the first 2 to 3 weeks, the better the prognosis. The cellular basis of this overlapping pattern of syndromes will be explained later in this chapter.

Radiation sickness

After a single dose of radiation above 50 rads, given to the whole body, symptoms may appear in one to two hours. The onset, duration and severity of all symptoms varies, depending largely on dose and partly on susceptibility. Symptoms may include (a) general: headaches, vertigo, debility and abnormal sensations of taste or smell; (b) gastro-intestinal; anorexia, nausea, vomiting, diarrhoea; (e) cardio-vascular: tachycardia, arrhythmia, hypotension and dyspnoea; (d) haematological: leucopenia thrombopenia and increased sedimentation rate; (e) psychological: increased irritability, insomnia and fear.

The incidence of radiation sickness is affected by the parts of the body irradiated. Exposure over the whole trunk and particularly over the upper abdomen causes more radiation sickness than does exposure of comparable tissue volumes in the extremities.

General clinical picture

Although the different organs have widely different radiosensitivities for the acute radiation syndrome in man, the three important systems are the central nervous system (CNS) particularly the brain; the gastrointestinal system, particularly the small intestine; and the haemopoietic system, i.e. the bone marrow together with lymphoid tissue. The acute radiation syndromes may therefore take three primary forms — cerebral, gastrointestinal and haemopoietic, in decreasing order of dose. To induce acute effects in the central nervous system requires several thousand rads; damage is then seen within minutes to hours. The dose for the acute small intestine form is more than 500 rads, with a latent period of several days. For severe haemopoietic changes, this is more than 200 rads and the effect takes about two weeks to develop (Fig. 9.1).

The clinical descriptions that follow are based upon subjects who suffered radiation accidents. The usual uncertainty of radiation dose distribution together with the variable effect of clinical treatment upon each individual may influence some details of the symptomatology. The qualitative descriptions probably remain valid but the quantitative information may be of less value. In particular, the time course of haemopoietic death might prove to be shorter than the following description, in the academic situation of a man exposed to total body radiation and subsequently not treated. This is indicated by extrapolation from the experimental animal studies which will be described later.

In each case the radiation dose level is assumed to be from one acute exposure. If protraction or fractionation of radiation dosage occurs then the usual

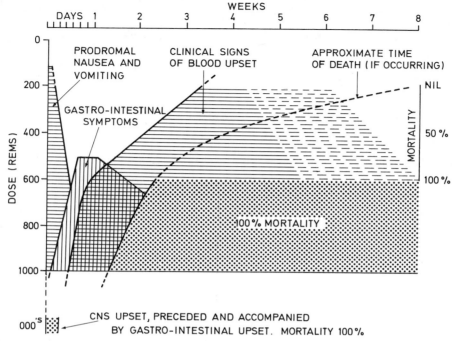

Fig. 9.1   Time sequence of the main events of the acute radiation syndrome in man (from Blakely, 1968).

biological consequences will apply, i.e. a correspondingly larger total dosage will be required to produce the same effect.

CNS Syndrome

After a single dose of several thousand rads to the whole body, in particular to the head, the clinical onset is prompt and death may occur in minutes to hours. After the initial phase of radiation sickness, there is swift progression from listlessness, drowsiness, and languor to severe apathy, prostration and lethargy, probably caused by small non-bacterial inflammatory foci appearing throughout the brain in one to two hours. This development of vasculitis or encephalitis gives rise to cerebral oedema. After more than 5000 rads, one deals with seizures ranging from generalized muscle tremor to epileptoid convulsions similar to grand mal. This convulsive phase lasts a few hours and is followed by ataxia from vestibulo-cerebellar disturbance. Convulsions and ataxia probably result from the degenerative pyknosis in the granule layer of the cerebellum within two hours after exposure, concomitant with brain oedema. Total body radiation causing the CNS syndrome is fatal.

Gastro-intestinal syndrome

The gastro-intestinal form predominates with lower doses received by the whole body, in particular to the abdomen (500 to 2000 rads). The prodromal nausea and vomiting begin promptly and do not subside. For some people, these symptoms develop within half an hour after exposure; in others, not for several hours. Gastrointestinal symptoms may continue (anorexia, nausea, vomiting, and diarrhoea). Sometimes the symptoms disappear after two to three days and recur by about the fifth day (just when the patient's condition seemed to have improved) owing to injury of intestinal epithelium, which by then is denuded of cells leaving few and even bare villi. (This cellular effect will be discussed below.) Rather abruptly, malaise, anorexia, nausea and vomiting prevents normal food and fluid intake, leading to serious electrolyte imbalance. Simultaneously, high fever and persistent diarrhoea — rapidly progressing from loose to watery, bloody stools — appear. The abdomen is distended and peristalsis is absent. Rapid deterioration leads to severe paralytic ileus. Exhaustion, fever and perhaps delirium follow; dehydration and haemoconcentration develop; the circulation fails, and the patient becomes comatose and dies, a week or two after exposure, with circulatory collapse.

After doses where regeneration of the gastrointestinal epithelium is possible antibiotics, fluid replacement, and other supportive therapy may keep the patient alive. The epithelium regenerates and vomiting and diarrhoea subside. This is only a temporary respite as evidence of bone marrow aplasia and pancytopenia begin within two to three weeks. After doses that cause this severe intestinal damage, bone marrow regeneration is unlikely, so that even if there is spontaneous recovery or successful treatment, individuals have yet to experience the effects on haematopoiesis.

Haemopoietic syndrome

In the haemopoietic form, after lower doses of irradiation, i.e. less than 500 rads, the haemopoietic symptoms are due to different origins and appear in two successive phases. Leucopenia, thrombocytopenia and haemostatic abnormalities are a direct consequence of lesions of the haematopoietic organs. Symptoms such as haemorrhage and anaemia may be secondary to the visceral lesions and associated with oedema or thrombosis in capillaries and ulceration of mucous membranes. Anorexia, apathy, nausea and vomiting, and some diarrhoea are maximum six to twelve hours after exposure. The symptoms may subside so that by twenty-four to thirty-six hours individuals feel well, but their bone marrow, spleen, and lymph nodes are atrophying. The patient enjoys apparently normal health until about the third week. Then chills, malaise, and fever, headache, fatigue, anorexia and dyspnoea on exertion develop, and at this time partial or complete loss of hair is likely.

Within a few days the general condition worsens and the patient then develops a sore throat and pharyngitis, accompanied by swelling of gingiva and tonsils, and petechiae in the skin with a tendency to bruise easily. This is followed by bleeding from gums and ulcerations on gingiva and tonsils. Similar ulceration in the intestines causes a renewal of diarrhoea. The patient has a high fever with complete anorexia. During weeks five to six, agranulocytosis, anaemia, and infection becomes critical. The increased susceptibility to infection is caused by the dose dependent decrease in circulating granulocytes and lymphocytes, (see 'discussion of cellular mechanisms' below) impairment of antibody production, impairment of granulocyte and reticulo-endothelial functions, and haemorrhagic ulceration permitting entrance of bacteria through the gastrointestinal tract. Thereafter, if the patient recovers, fever, petechiae and ecchymoses subside; ulcerations heal and convalescence begins about the end of the second month after exposure.

Prognosis. A knowledge of these symptoms and signs allows a prognosis to be determined in those cases of accidental total body irradiation where there is considerable uncertainty as to the dose received by the patient. Individuals showing the CNS syndrome will die. Subjects with intractable nausea, vomiting, and diarrhoea will generally die. Those in whom nausea and vomiting is brief, i.e. for one to two days followed by wellbeing, have a good chance of survival.

After initial symptoms, the effects of haemopoietic damage predominate. The lymphocyte count is valuable as an early criterion for judging radiation injury. In normal individuals, a fall in lymphocyte number is seen within the first twenty-four to forty-eight hours. If, at forty-eight hours, the lymphocyte count is 1200 or above, it is unlikely that the individual has received a fatal exposure; if the lymphocyte count is in the 300 to 1200 range, a dose in the lethal range may be suspected; counts below 300 indicate an extremely serious exposure. (These counts would presumably be raised if any cells have been transfused.)

The total white cell count is of particular value for following the patient throughout the course. In general, the drop in neutrophils will reflect the degree of exposure; a fall in white counts beinning within the first week denotes a rather high exposure, whereas late falls indicate a less serious exposure. Because of this time course of the total blood count and the various elements, a consideration of the acute radiation syndrome, especially when it is followed by death, must include a discussion of the time intervals involved. A consideration of the concept of LD_{50} (mean lethal dose) will illustrate this.

Mean lethal dose – LD_{50}. Figure 9.2 illustrates a lethality curve for Rhesus monkeys and it is seen that a dose of 528 rads is quoted as the LD_{50}. It can be seen from this curve that this is a statistical statement implying that in a large enough sample of animals 50 per cent would be killed by a dose of 528 rads to the whole body. It is important to note, however, that many animals will survive a dose of 550 rads, some even of 600 rads, and also that some animals may die from a dose of as low as 450 rads. Nevertheless, the dosage range in Figure 9.2 is comparatively narrow, i.e. the survival curve has a broad shoulder after which there is a rapid drop in the fraction of animals which survive. This implies a threshold level of survival of some critical group(s) of cells, a concept which will be discussed later in this chapter (the cellular basis of the acute radiation syndrome).

The LD_{50} is a useful statement when comparing different animals or different methods of delivering the total body radiation, or perhaps of treating the animals after this radiation; but it is still only a guide as far as any individual

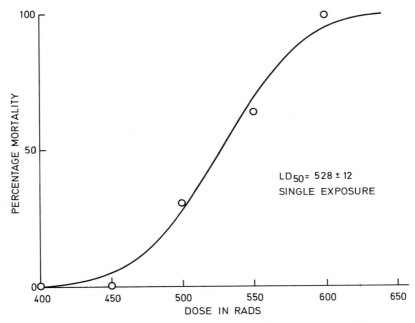

Fig. 9.2 Lethality curve for rhesus monkeys (from Paterson, 1954).

animal would be concerned when assessing prognosis. The expression LD_{50} is thus a convenient way of smoothing out the problems of heterogeneous populations; smoothing out the results which would be obtained from unduly sensitive or unduly resistant subjects. LD_{50} values are known for a number of animals (Bond and Robertson, 1957) but the value for man can only be estimated; it is shown as 400 rads in Figure 9.1.

There is a further definition that must be added to the expression LD_{50}, however, in order to assess the type of radiation death. Table 9.1 illustrates the

Table 9.1 Dose (rads) of total body radiation and time of death

Species	$LD_{50/30}$	$LD_{50/8}$	$LD_{50/5}$	$LD_{50/2}$
Mouse	640		1260	20 000
Germ-Free Mouse	705	2000		
Rat	714		808	20 000
Monkey	600	1500		10 000

(Data from Bond, Fliedner and Archambeau, 1965)

dosages which will kill 50 per cent of a sample of different animals and in this Table a subscript has been added to the expression LD_{50}. Under these different headings different dosages of total body radiation are required to kill 50 per cent of the animals. This subscript is the number of days at which the number of deaths is assessed. In the left-hand column, $LD_{50/30}$, a considerable time, 30 days, has elapsed since the dose of total body radiation and this is usually regarded as sufficient for the final results of the syndrome to be established. This is the most convenient form of the expression LD_{50} and when no subscript is used it is usually assumed that this is the expression, although the corresponding time interval for man is 60 days.

The other columns give the LD_{50} values for shorter time intervals and, for these, higher dosages of radiation are quoted. The right-hand column, $LD_{50/2}$, is death at two days after the dose of radiation and it should be obvious that this must be death from the CNS syndrome, which has already been discussed. The choice of two days as a time limit reflects the observations that animals dying prior to two days show few, if any gastrointestinal signs and symptoms, whereas those animals dying later than this time, do. It is possible or perhaps probable that deaths attributed to the CNS syndrome on this basis actually include a variety of causes of death that have not, as yet, been separated out.

The choice of the time limit of 5 (or 8) days reflects the tendency of deaths in the transition zones to group themselves around two, sometimes overlapping modes at less than 5 days (or 8 days). Associated with this distribution there is a correspondence of the clinical course, with signs and symptoms of the gastrointestinal syndrome predominating at the shorter survival times. This is not to say that the bone marrow syndrome may not already have become evident, but it is the mode of death which is under consideration here. Just as it was CNS death at a much earlier phase, it is now the gastrointestinal syndrome which is causing death in this phase.

Death of animals from the bone marrow syndrome occurs commonly at 12 to 14 days. This time interval is well outside the 8 day period for gut death but since there is not a fourth mode of death from acute radiation to follow that of the bone marrow syndrome, the interval of 30 days has been accepted. 30 days is a convenient experimental period after which animals may be considered to be survivors from acute radiation. Subsequent deaths can reasonably be excluded from an acute radiation cause.

The three expressions $LD_{50/2}$, $LD_{50/8 \text{ or } 5}$ and $LD_{50/30}$ are statistically convenient, but deaths from acute radiation cannot always be divided into separate categories. The three syndromes may often co-exist. Nevertheless, the three modes of death will again be classified on this basis to enable a cellular explanation of these syndromes.

The cellular basis of the acute radiation syndromes

This section will develop the thesis that the nature of radiation death, the timing of this death and whether death occurs at all, or whether recovery ensures, can be shown to depend upon various cell populations, their depletion, and whether they are able to recover to a normal level. In fact, the main discussion must be confined to the second two modes of death — gastrointestinal and haematopoietic — because, although a cellular explanation might be found in time to apply to CNS death, in practice the effects are so acute that there is not time for this to become manifest.

Pathogenesis of CNS death

The findings in CNS death are vasculitis and meningitis with cerebral oedema. The neurons of the cerebral cortex show few, if any, morphological changes. There is ample evidence of vessel and capillary damage following high dose head or whole-body radiation but the cause of death is not known with certainty. Presumably death results from events that are evolving within the confined limits of the skull; however, it is not clear if death is a result of irradiating neurons *per se,* or the blood vessels, or both. In any event, death presumably results from neuronal change, either directly or via increases in intracranial pressure.

When discussing the immediate causes of death in the CNS syndrome it is important to remember that neuronal tissue is a non-proliferating cell population and demonstrates fewer early effects of radiation than a proliferating one. When tritiated cytidine is given to head-irradiated rats, there is continued, though depressed, uptake in the neuronal tissue, which indicates that metabolic function continues in these cells. This is suggestive evidence that neuronal tissue as such is not killed outright by these high doses of radiation, and is confirmed by the fact that, following partial radiation of the brain, necrotic neuronal changes require several weeks to evolve. Although the damage to the cells may be lethal, it would appear not to be acutely so. Thus, the mechanism of death following doses of radiation in excess of the threshold dose of the CNS syndrome, would appear to be neuronal damage secondary to vascular damage, oedema, and increased intracranial pressure.

Cell population kinetics

This section will consider the kinetics of those cell populations which do proliferate, namely, the gastrointestinal epithelium and the haemopoietic system. The radiation response of both gut and bone marrow were discussed individually in Chapter 8, with other normal tissues. In the context of the radiation response of the total body, however, the kinetics of the two stem cell populations merit further study. Figure 9.3 shows a theroretical diagram of the kinetics of these various cellular systems. The diagram follows, over a course of time, the percentage of viable cells after a dose of X-rays delivered at X. It is known (and this has been discussed in earlier chapters) that the larger the dose of X-rays the smaller the surviving fraction of viable cells. The end result of a small or a large dose of radiation is shown in the diagram. The time scale can be days, or whatever is appropriate to the particular cell population; dependent upon the length of the cell cycle. During the first few days, the number of cells falls at a steady rate until the surviving fraction of cells has regenerated sufficiently to begin repopulation, when the curve begins to rise again. A level of risk can be defined, below which the integrity of the tissue and therefore the viability of the whole animal may be at risk. Thus the size of dose and therefore the extent of the cellular depletion, and the rate of its recovery (either the slower rate of the solid lines or the faster rate of the dashed line) may influence

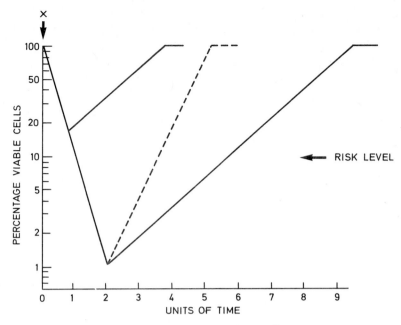

Fig. 9.3 Kinetics of irradiated cells.

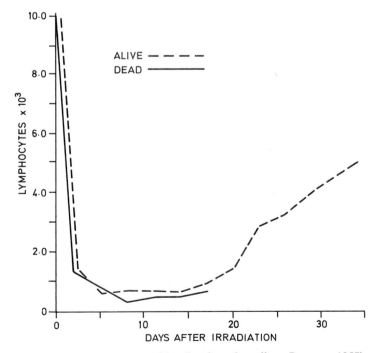

Fig. 9.4 Lymphocyte counts of irradiated monkeys (from Paterson, 1957).

the period of time during which the animal is at risk with respect to a damaged cellular system. An actual example of this is shown in Figure 9.4 where the lymphocyte count is shown in two monkeys, given total-body radiation. In both cases the count fell to a very low level, from which one of the monkeys recovered and the other did not. Another example is taken from mice in Figure 9.5 which is a schematic diagram of the amount of sperm available for ejaculation. Here the same shape of curve applies, although in this case the period of risk does not involve danger of death but merely a period of sterility (see Ch. 11).

Cell renewal systems

Figure 9.6 shows a model, the cell renewal system, which in this case is applicable to the myeloid cellular system, but can be applied to any proliferative cell population. It is assumed that there is a pool of stem cells which have survived the dose of radiation and that, after a given *transit time*, this pool may replenish itself and then replenish the peripheral system that is required for the integrity of the tissue and of the animal as a whole. It can be seen that a time period is necessary before the new cells will become available to the tissue from the surviving stem cells, but that during this time the already mature cells will remain viable for a period depending upon their life-span. It is the balance between the gradual removal of the ageing mature cells and the arrival of the new mature cells that will determine whether the tissue remains intact to a sufficient degree to maintain the integrity of that tissue and therefore the life of the animal.

An example of the myeloid system of a monkey, given total-body radiation, is shown in Figure 9.7 of the polymorphonuclear white-cell count of such an animal which recovered. This figure shows, as did the previous one illustrating the lymphocyte counts, that the critical time period for such an animal is from 12 to 14 days, and this illustrates why animals dying haemopoietic deaths will die during this time. This is the period of risk of the haemopoietic system, the period when the peripheral blood levels are at their lowest and when the animal is most susceptible to all the manifestations of haemopoietic inadequacy. The bone marrow may well be in active regeneration already, by this time, and the animal might well have the capacity to recover eventually if the peripheral blood elements can somehow be replaced, (e.g. by transfusion of packed cells) for a long enough period to allow regeneration of the bone marrow to manifest itself in the peripheral blood, as seen in these diagrams.

From this specific example of the cellular basis of haemopoietic death, it is possible to generalize to other cellular systems. It is only the time scale that requires to be altered. The kinetics of a depression in cell number followed by recovery will follow the same general shape shown in our original Figure 9.3. This relationship was illustrated in Chapter 3, Figure 3.1 where the transit times of a number of cellular systems were compared. There was a constant relationship between the *transit time* of non-proliferating maturing cells (the cells that are going to form the peripheral part of the system) and the days' period of time from irradiation before the previous mature cells were depre-

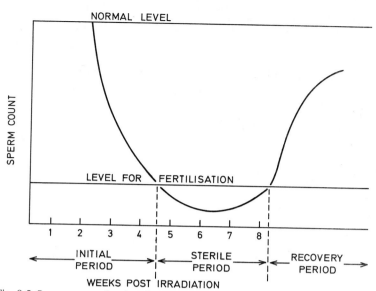

Fig. 9.5 Sperm count after irradiation (from Paterson, personal communication).

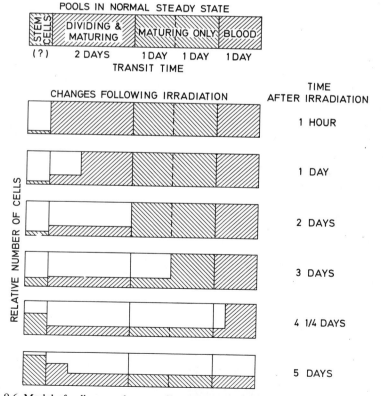

Fig. 9.6 Model of cell renewal system (from Bond, Fliedner and Archambeau, 1965).

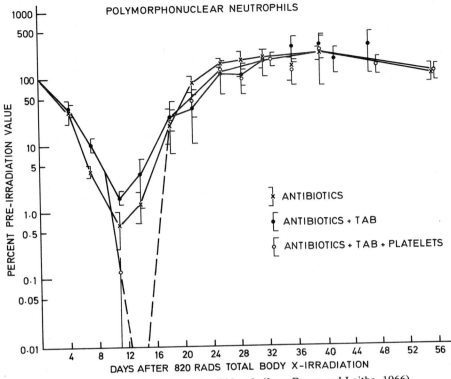

Fig. 9.7 Polymorph count of monkeys after 820 rads (from Byron and Lajtha, 1966).

ssed in number. The earliest cellular system in that figure was the intestine, the bone marrow systems came later, and finally, the testis. This is illustrated in another way in Figure 9.8 where the depression of several cell populations is shown over a certain number of days. (These cell populations were discussed in Ch. 8.)

It is because of these different cellular responses that the various modes of death of the acute radiation syndrome follow the time course already discussed. The reason that gastrointestinal death occurs before haemopoietic death is due to the faster depression in the gastrointestinal epithelial cell population. This is due to its shorter transit time and is illustrated in Figure 9.9 where the rapid fall in intestinal epithelial cell count reaches a level where diarrhoea ensues. Figure 9.9 also illustrates the delay in this fall in cell number which occurs in animals which have been kept germ free. Death from the gastrointestinal syndrome would also be delayed for this period in such animals. This is why Table 9.1 quotes the LD_{50} figure for germ-free mice under the $LD_{50/8}$ column rather than the $LD_{50/5}$ column.

It should now be obvious why Table 9.1 is set out in this form. It has already been mentioned why CNS death is such an acute manifestation: because of immediate vascular effects ($LD_{50/2}$). The middle columns of LD_{50} are dealing predominently with the gastrointestinal syndrome which occurs before five to

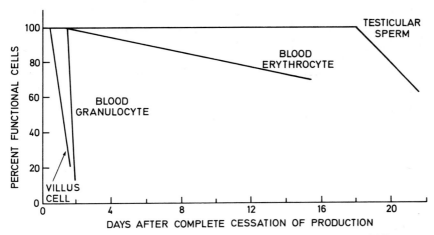

Fig. 9.8 Time intervals of response of normal cell populations (from Patt, 1963).

eight days because of the time course of the depletion of the gastrointestinal epithelium which falls below a risk level before five to eight days. If death from this syndrome is to occur it will occur in this period of time, otherwise the epithelial system will already be regenerating and the animal will recover as far as this system is concerned. Furthermore it can be seen that the haemopoietic system, as far as its functional peripheral compartment is concerned, has still to reach its time of greatest risk which may occur between day 10 and 20. For this reason, the first column in Table 9.1 is expressed as $LD_{50/30}$ days in order to allow for death to occur during this thirty day period.

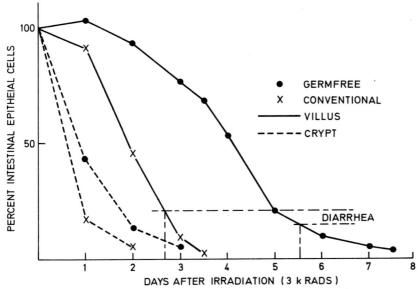

Fig. 9.9 Depression of intestinal epithelial cell counts (from Bond, Fliedner and Archambeau, 1965).

These principles have been described for laboratory animals which have provided accurate information not normally available for man. The same principles can be applied to man but with a longer time span, as was shown earlier in Figure 9.1. This also showed that the various acute radiation syndromes occur over a range of times which overlap each other. The syndromes have only been described separately for convenience.

Partial body radiation

The same cellular basis of acute radiation can be used for the consideration of partial body-radiation. If it is known which part of the body has been irradiated, one can consider which cell population may be at risk. Table 9.2

Table 9.2 Dose of partial body radiation and time of death

Rat	$LD_{50/30}$	Mean Survival Time
Upper Body	1800 rads	11 days
Lower Body	1000 rads	5 days
(abdomen)		

(Data from Bond, Fliedner and Archambeau, 1965)

compares the $LD_{50/30}$ values of rats treated either through the upper body or through the lower body. There is a considerable difference between the two figures and both figures are higher than that shown for the $LD_{50/30}$ of the total-body in Table 9.1. Much less tissue is at risk but the particular tissue at risk still influences the lethal effects. With the lower body, it is predominantly the gastrointestinal tract which is affected and death occurs at an average of five days, as is usual with the gastrointestinal syndrome; although a higher dose is tolerated because of the smaller amount of tissue involved. When only the upper part of the body is irradiated, an even higher dose is tolerated and the mean survival time rises to eleven days; more typical of the haemopoietic death. This is related to the incidence of the irradiation on the bone marrow, chiefly present in the upper body. But much bone marrow is still spared so that the LD_{50} figure is considerably higher than that for the whole body. Thus, although the LD_{50} figures are higher, the same cellular basis can be applied and the time course of death may remain the same. Large doses to the whole thorax may also result in early death due to oesophageal damage. (This will be discussed in Chapter 11 in contrast with the 'late effect' of lung fibrosis which may also follow thoracic irradiation.)

When even smaller volumes than the upper or lower halves of the body are involved, one is concerned with tissues, which, although when damaged may lead ultimately to death, will not lead to the acute death that has been considered so far. Some tissues, indeed, may be considerably damaged without causing death at all, although serious functional defects will be suffered by the animal for the rest of its life, and possibly, this life-span may be shortened. These effects will be discussed in Chapter 11.

REFERENCES

Blakely, J. (1968) *The Care of Radiation Casualties,* London: Heinemann.

Bond, V. P. & Robertson J. S. (1957) Vertebrate radiobiology (lethal actions and associated effects). *Annual Review of Nuclear Science,* **7**, 135-162.

Bond, V. P., Fliedner, T. M. & Archambeau, J. O. (1965) *Mammalian Radiation Lethality.* New York. Academic Press.

Byron, J. W. & Lajtha, L. G. (1966) Radioprotection in total-body irradiated primates. *British Journal of Radiology,* **39**, 382-385.

Lushbough, C. C., Comas, F. & Hofstra, R. (1967) Clinical studies of radiation effects in man. *Radiation Research supplement,* **7**, 398-412.

Paterson, E. (1954) Factors influencing recovery after whole-body radiation. *Journal of the Faculty of Radiologists,* **5**, 189-199.

Paterson, E. (1957) The mechanism of death following whole body irradiation. *British Journal of Radiology,* **30**, 577-581.

Paterson, R. (1963) *The Treatment of Malignant Disease by Radiotherapy.* London: Edward Arnold.

Patt, H. M. (1963) Quantitative aspects of radiation effects at the tissue and tumour level. *American Journal of Roentgenology,* **90**, 928-937.

Report of the United Nations Scientific Committee on The Effects of Atomic Radiation, 1962, New York, United Nations.

10. Response of Tumours

It has already been described how the inevitable result of exposure to ionising radiation is the destruction of biological material in the treated area. Earlier chapters have dealt with the immediate response of normal tissues which limit the tumour dose, either because of an upper limit of early damage to particular tissues (like skin, blood vessels, gut or bone marrow; Ch. 8) or because of a limitation when large volumes of the body are included in the radiation field (Ch. 9). The dose delivered to a tumour will also be limited by the risk of producing late damage which becomes manifest in certain organs (like lung, kidney, thyroid, CNS, and the lens of the eye, Ch. 11). The control of malignant disease remains the prime object of radiotherapy, however, and this present chapter is devoted to the radiation response of tumours. The emphasis will be upon scientific laboratory observations of the response of experimental animal tumours to radiation, as distinct to those clinical observations (e.g. from controlled clinical trials) which are usually less well defined and relatively imprecise.

Kinetics of tumours

Experimental radiotherapy should be performed on those animal tumours which are reasonable models of human cancers. Histological classification into carcinomas, sarcomas, lymphomas, etc. can be related to animal models (although most experimental animal tumours grow much faster than human tumours). The growth characteristics of some human tumour cell populations were discussed in Chapter 5 and examples of the kinetic parameters were given using the growth fraction model. (Fig. 5.7). In this, tumour cell populations are shown to have four compartments: cells in cycle with a measurable cycle time; resting cells (or the *non*-growth fraction) which may still have the capacity to grow; sterile cells which are still alive but have lost the capacity to reproduce; and dead cells, which eventually will be 'lost' from the population.

Table 10.1 lists five histological types of human tumour with the average

Table 10.1 Kinetics of human tumours

Histological Type	Number of Tumours	Doubling Time (days)	Growth Fraction %	Cell Loss %
Embryonal	6	27	90	94
Reticulosis	15	29	90	94
Sarcoma	32	41	11	68
Squamous Cell Carconoma	68	58	25	90
Adenocarcinoma	121	83	6	71

(from Malaise, Chavaudra and Tubiana, 1973)

values for growth fraction, cell loss factor and doubling time. Although this table comprises observations from 242 human tumours, some of the histological types are poorly represented. Nevertheless, the parameters provide some basis for comparison with the animal tumours which have been used for radiobiological studies. For convenience, the tumours are listed in increasing order of doubling time but there are variations in the values for cell loss factor, and particularly for growth fraction, which account for the fact that the cells in such tumours are often cycling at a very much faster rate than the doubling time would suggest. Thus, for squamous cell carcinoma, labelling studies indicate a cycle time of about 6 days. The volume doubling time is very much longer at 58 days because only 25 per cent of the tumour cells are in the growth 'compartment' and 90 per cent of their surplus progeny are 'lost'.

Table 10.2 lists four of the many animal tumour systems which have been used in radiobiology. They serve to illustrate the problem of finding a suitable

Table 10.2 Kinetics of animal tumours

Animal	Code	Histological Type	Doubling Time (hours)	Growth Fraction %	Cell Loss %
Mouse	L1210	Leukaemia	10	95	5
Rat	R1B5	Fibrosarcoma	24	45	0
Rat	R1	Rhabdomyosarcoma	66	29	62
Mouse	C_3H	Mammary Carcinoma	110	30	70

'model' tumour. The L1210 leukaemia was described in Chapter 6 (Fig. 6.3). This system has been widely used for the screening of cancer chemotherapy compounds but the very low cell loss factor shows it to be quite unlike any of the human tumours. The same criticism can be applied to the R1B5 fibrosarcoma. Some of the radiobiological findings with this tumour will nevertheless be quoted in this chapter because they illustrate some important general principles. The more 'relevant' tumours are the R1 Rhabdomyosarcoma (Fig. 10.4) and the C_3H mammary carcinoma. These two are examples of tumours with a high cell loss factor (60 to 70 per cent) and a growth fraction of the same order of magnitude as some of the more common human tumours (less than 30 per cent).

Reference will be made to many other 'model tumours' which may be grown and transplanted in small animals. Small animals are used for convenience of laboratory space and because the strictly controlled breeding of strains of mice and rats enable the radiobiologist to perform studies on large numbers of similar tumours under constant and reproducible conditions to a high scientific standard. Tumours can be transplanted into groups of animals which are identical genetically (isologous) so that the only variable in the experiment will be the radiobiological factor to be tested. There remains one important difference between mice and men, however. This is the growth rate of the tumours. Many of the murine tumours have a doubling time of 2 to 3 days whereas many human tumours have a doubling time around 70 days. This large discrepancy might be accounted for by comparing the normal life span of mice and men; namely 2 to 3 years and 70 years respectively.

Unfortunately the growth kinetics of the limiting normal tissues of the two species do not show anything like that discrepancy (see Ch. 8). Furthermore many experimental tumours in the mouse are allowed to grow to a relatively very large size. A one gramme tumour in a mouse would be equivalent to a 3 Kg tumour in man, which would be enormous by any standard. With such large tumours the mice often develop anaemia, with the haemoglobin level falling below 10 per cent. This is associated with an increase in the fraction of hypoxic cells to 21 per cent but this can be reduced again to about 8 per cent by a transfusion of packed red cells.

The only solution to this problem of comparison is to study tumours of a small size (say 5 mm diameter, 0.2 g in mice). Each experiment should preferably be carried out on a range of animal tumours; chosen to cover a range of variables which show similarity to the human situation in at least one respect, since no exact matching can be achieved. As long as the animal tumours are used as *models* for the purpose of examining specific effects and responses then the tremendous advantage of obtaining statistically significant results will outweigh the disadvantages of the differences in the biological characterisation of tumours in mice and men.

Growth restraint

Growth restraint is one of the objectives of palliative radiotherapy and it can be studied with an animal tumour which shows a poor response to radiation. (It will be noted that the word *radioresponsive* is usually preferred to 'radiosensitive'.) The R1B5 fibrosarcoma can be transplanted into Wistar rats and the growth of the transplants can be measured in terms of the diameter of the tumour as shown in curve A of Figure 10.1. The vertical scale in this figure is a linear one of mean diameter. In other graphs of tumour growth the scale may be logarithmic with volume as the measurement. An estimate of volume may be derived from diameter, and volume doubling time calculated in this way is the parameter used in Tables 10.1 and 10.2. Most tumours do not show an exponential growth curve during their whole life-span. A semi-logarithmic plot for the R1 Rhabdomyosarcoma in rat is shown in Figure 10.4 and this is bending towards the time axis, indicating a slowing in volume doubling time throughout the whole series of measurements. This is probably due to deteriorating nutritional status as the distance between blood vessels increases and widespread local necrosis results.

Detailed kinetic studies now show that there is a decreasing gradient in doubling times between cells in the centre and at the periphery of larger tumours, in addition to the longer doubling times of larger tumours in general. It is necessary always to study tumour growth-rate data with care. Are they mean values from the whole tumour? Are they mean values of the whole life-span of the tumour from early to late? Are they expressed as a logarithmic plot of volume? How was this volume measured; indirectly by weighing, or by calculation from records of the diameter of one or more of the three dimensions of a tumour? A well documented radiotherapy case history should contain a series of measurements of tumour diameter. Most of these must necessarily follow treatment but the sort of pattern illustrated for the rat Rhabdomyosar-coma in Figure 10.4, and for the Fibrosarcoma in Figure 10.1 in this chapter can be recognized. For ethical reasons those parts of the two figures which show tumour growth before or without treatment will usually remain absent from clinical records. But these can be derived by extrapolation, on the assumption that these animal models apply to human tumours.

Figure 10.1 shows (on a linear scale of diameter) how increasing single doses of radiation leads to growth restraint for increasing periods of time (A–G

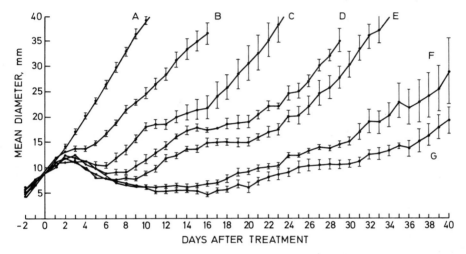

Fig. 10.1 Growth cruves of rat fibrosarcoma (from Thomlinson, 1961).

corresponds to 0, 1000, 2000, 3000, 4000, 5000 and 6000 rads). If the sur-rounding tissues could only tolerate an even larger dose then this period of growth restraint would become so long that 'cure' would have been achieved. More commonly, residual cells grow and the tumour regenerates at a rate which may be similar to that before treatment. On the linear plot in Figure 10.1, however, the post-treatment curves are not parallel to that of the untreated tumour. This may reflect a real change in the growth rate of the surviving tissue. The data from human tumours are too scanty to decide the issue, but the semi-logarithmic plots of animal tumour data suggest that no change in growth rate occurs.

Uniform growth patterns before and after irradiation are illustrated in Figure 10.2 where serial measurements of volume of a transplantable osteosar-coma in the mouse are plotted on a semi-logarithmic scale. Over this range of volumes, the untreated tumour shows approximately an exponential growth pattern and the final slopes of the recurrent tumours can be drawn roughly parallel to that. This figure shows the use of an animal tumour for the study of fractionated radiotherapy, and chemotherapy with cyclophosphamide, either four or five times a week over a three week period. While five treatments of cyclophosphamide per week produced more growth restraint than four there

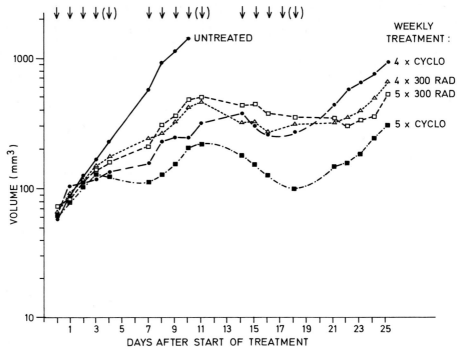

Fig. 10.2 Growth curves of mouse osteosarcoma (from Van Putten, 1969).

was very little difference between the results of the two radiation regimes. This suggested that the fifth radiation dose each week was less effective than the four other treatments and it was considered that this was because this particular tumour is slow to reoxygenate. Thus the fifth radiation dose each week was relatively ineffective because of the increased proportion of relatively resistant hypoxic cells. A longer 'week-end's rest' would permit more reoxygenation so that the dose on Monday would be more effective than the treatment on a Friday. This leads to a fuller consideration of the vascular supply to tumours (discussed in Ch. 3 and 8) and their oxygenation.

The vascular supply of a tissue can be studied by measuring the rate of clearance of a labelled inert gas like Krypton or Xenon which will be respired from the pulmonary vessels. Very small volumes of [133]Xenon can be injected directly into selected sites of a tumour and a comparable normal tissue. The radioactivity of the injected volume is measured by placing a collimated scintillation counter over the site of injection and the rate of blood flow can then be calculated from the clearance half time. This technique has been used in human as well as animal tumours and the rate of clearance is found to decrease with increasing tumour volume, while normal tissues maintain a fast clearance rate. While this measurement of the clearance of interstitially deposited [133]Xenon remains an indirect indicator of blood flow in a tissue, it does permit study of the effects of radiation on the vascular supply of tumours and thus, indirectly, on their state of oxygenation.

Reoxygenation

Reference has already been made to the influence of anaemia on the proportion of hypoxic cells in a tumour. Radiotherapists usually maintain the haemoglobin level of their patients above the 70 per cent level for this reason. Even then, the majority of tumours are believed to have regions where the vascular supply is inadequate so that even a patient with a normal haemoglobin level will be unable to maintain a physiological oxygen tension throughout the cells of his tumour. Figure 10.3 is the classical diagram which shows the relatively small

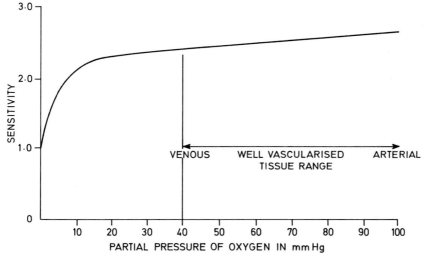

Fig. 10.3 Relationship of radiosensitivity to oxygen concentration (redrawn from Deschner and Gray, 1959).

fall in radiosensitivity of tissues nourished by a normal blood supply, i.e. the usual gradient from arterial to venous blood. Diffusion studies suggest that oxygen cannot diffuse more than 150μm from capillaries so that any tissue with an inter-capillary distance of more than 300μm will contain anoxic cells. Other chemicals have different diffusion rates and the value of the new class of *electron-affinic compounds* (Ch. 2) is that they can diffuse further than oxygen and are not metabolised in the hypoxic human tumours which they are intended to sensitise.

Figure 10.3 shows how the radiosensitivity of biological material falls rapidly when the partial pressure of oxygen is decreased below 20 mmHg; well outside the physiological range. It is assumed that such a phenomenon does not occur in most normal tissues but will be found in tumours. This assumption needs to be modified to take into account those tissues, like brain, where capillaries are known to undergo cycles of constriction and dilatation. The time scale of those cycles is usually measured in minutes, whereas the undoubtedly dynamic state of a tumour blood supply is more likely to be measured in cycles of hours. For the purposes of fractionated radiotherapy, delivered at a normal dose-rate, the simplified assumption that normal tissue *is* well oxygenated whilst a proportion of tumour cells is *not,* would seem to be valid.

In Figure 10.3 the relative sensitivity to low LET radiation is shown to vary by a factor 2 to 3; this is the *oxygen enhancement ratio.* Examples were quoted in Chapter 6 for several cell systems and the limiting normal tissues in Chapter 8 showing that this size of oxygen enhancement ratio is generally applicable. Many of the tumours studied in animals as transplantable tissues can also be examined in tissue culture so that accurate values have been obtained for OER. One example, only, will be quoted. The R1 Rhabdomyosarcoma, was one of the first animal tumour models to be investigated in this way.

Figure 10.4 combines (A) observations on the tumour measured as it grows in the animal and (B) its cells assayed *in vitro.* The shape of the upper growth curve (1) of the untreated tumour was discussed in Chapter 5 (Fig. 5.8). The lower growth curve (2) shows the response of the tumour to a single dose of 2000 rads. The tumour first stops growing and then shrinks, but by the twelfth day after treatment growth resumes at a rate similar to that of the untreated tumour. During this period the viability of cells taken from the treated tumour was assayed by a clonogenic test *in vitro.* The results of this are shown in the lowest curve in Figure 10.4. The fraction of clonogenic cells falls to 10^{-2} (or 1 per cent) immediately after irradiation and remains at this level during the four day period when the tumour has stopped growing in the animal. Then the clonogenic fraction begins to rise but the tumour shrinks because of death and removal of that 99 per cent of the cells whose viability was destroyed by the irradiation dose. By the twelfth day, however, the clonogenic fraction has returned to unity, all the cells in the tumour are now viable and the tumour resumes that growth rate which would be expected from the other measurements. This *in vitro* assay provides a clear explanation of the radiation response of the tumour *in vivo.* The *in vitro* assay can also be used to test cellular radiosensitivity in terms of survival curves. The survival curves obtained in air and under hypoxic conditions have D_0 values of 120 rads and 295 rads respectively, so that the OER is 2.46, a typical value.

That is an over simplification of the oxygen effect in a tumour cell population which almost always contains a proportion of hypoxic and aerated cells. In the case of the Rhabdomyosarcoma 15 per cent of the tumour cells are estimated to be hypoxic. It follows that the survival curve for such tumour cells will be biphasic in shape, with the change from a steeper aerated slope to a flatter hypoxic slope occurring at a dose level which depends upon the proportion of hypoxic cells in the tumour at the time of irradiation and also on the slope of the curve for aerated cells. Single dose survival curves with such a shape were discussed in Chapter 6 (Fig. 6.6) but such curves only measure the response of a tumour cell population to a 'first' dose.

Changes in the oxygenation of a tumour during fractionated radiotherapy were assumed to explain the results shown in Figure 10.2. A more direct study is needed, however, and Figure 10.5 shows the changes in oxygenation with time in three tumours which are being discussed in this chapter. In each case a 'priming' dose of 1500 rads was given first and then the tumours were tested for their radiobiological response to a second dose delivered at increasing intervals

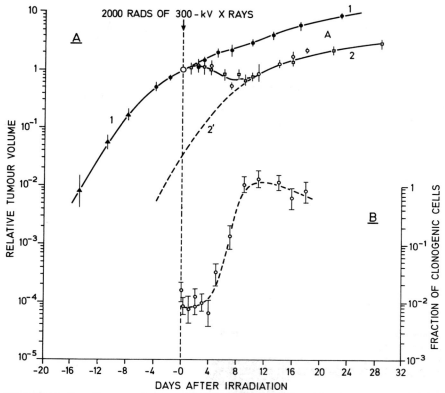

Fig. 10.4 A. Growth curves of a transplanted rhabdomyosarcoma (1) without treatment (2) after 2000 rads; measured *in vivo* B. Survival of cells from the irradiated tumour assayed *in vitro* (from Hermens and Barendsen 1969).

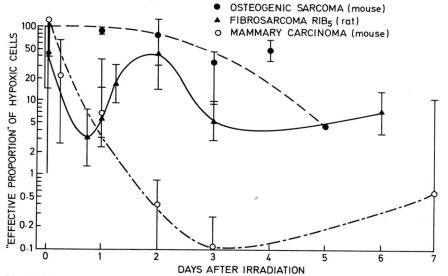

Fig. 10.5 Proportion of hypoxic cells in animal tumours (from Thomlinson, 1969).

after that first dose. The radiobiological response in terms of cell survival indicated the effective proportion of hypoxic cells in that tumour at that time. All three tumours start with a very high proportion, as would be expected after a dose of 1500 rads which would depopulate the tumour of the majority of its aerated cells. Thereafter the three tumours show different patterns: (1) Reoxygenation in the R1B5 brings the level down to 3 per cent within 24 hours but a 'hypoxic wave' occurs over the next 2 days because this sarcoma does not shrink until the second day after irradiation — it continues to enlarge (see Fig. 10.1). The second wave of reoxygenation occurring then can be attributed to shrinkage; the earlier wave to a greater availability of oxygen due to reduced respiration just after irradiation. (2) The C_3H mammary carcinoma falls to the much lower level of 0.1 per cent but not until 3 days. This is because the C_3H carcinomas do shrink at once (see Fig. 10.8), so the two waves of reoxygenation take place one after the other additively, and reoxygenation becomes much more complete than at any time in R1B5. (3) The proportion of hypoxic cells in the osteosarcoma has only fallen to 5 per cent by the fifth day.

The mechanism of re-oxygenation is ill understood but several factors may play a part. The death of some of the tumour cell population will result in a reduction in oxygen consumption, and the removal of dead cells may lessen the average intercapillary distance. Loss of tumour substance may help to improve blood flow because of reduction in tissue tension. An increase in vascularization may also occur. The temporal pattern and the degree of reoxygenation was shown to vary from one type of animal tumour to another (Figure 10.5) but it is difficult to obtain comparable information for tumours in man. If animal studies reflect the response of human tumours to radiation then the time honoured fractionation schedule of daily treatments, five days a week, may not prove to be the most effective way of delivering low LET radiation in cancer therapy. The influence of reoxygenation upon fractionation schedules will be further discussed in Chapter 13.

Tumour cure

Since the aim of radiotherapy is to cure the primary human tumour it is useful to study an animal tumour where this end can be achieved. In clinical parlance the word 'cure' is usually replaced by a statistical parameter like 5-year survival. For the mouse an equivalent period of time is 120 or 150 days and this is used in the studies shown in Figure 10.6. This is also part of a study of reoxygenation but this time the figure shows the results of an experiment using C_3H mice into whose chest wall the mammary tumour (which occurs spontaneously in this strain of mouse) has been transplanted. When the tumour measures 6 mm in diameter the experiment is begun with a 'priming' dose of 1500 rads. At time intervals thereafter the dose which cures 50 per cent of a group of animals is determined (TCD_{50}). In this case that second dose was given when the tumour was hypoxic (using a clamp around the tumour), or aerated or hyperbaric (with the mouse in a hyperbaric chamber). 'Cure' meant that there was not palpable evidence of tumour in the irradiated volume at 150 days after treatment.

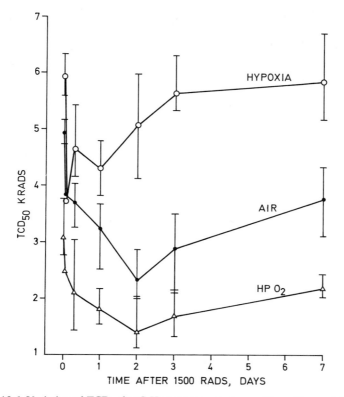

Fig. 10.6 Variation of TCD_{50} for C_3H mammary carcinoma (from Howes, 1969).

This is the most 'relevant' animal model discussed so far, although the dose needed to cure 50 per cent of animals in the aerated group (a total of 4000 rads with 2 days interval) may seem to be large by human standards. This may be because of the large value for D_o (>300 rads) found for C_3H mammary tumour cell survival curves, both *in vitro* and *in vivo*. Nearly all the animal tumour experiments described in this chapter require the use of a large number of animals but the sort of result shown in Figure 10.6 can then be obtained within a period of 4 to 6 months. Experiments in scientific laboratories must always be repeated but still the investigation can be completed within a year. This compares with the ten year period often required for a controlled clinical trial where 5 year survival is measured. The statistical significance of results like those in Figure 10.6 will also be superior to those from a clinical trial because the biological material is less variable. Thus hyperbaric oxygen is shown to sensitise the tumour to radiation and hypoxia to protect. In addition the reoxygenation observations for the C_3H tumour described earlier in Figure 10.5 are confirmed in Figure 10.6 for air and hyperbaric oxygen in that the tumour is most sensitive to a second dose delivered 2 to 3 days after a priming dose. This best interval for fractionation has also been shown to be an optimum interval for multifraction X-ray treatments of the same C_3H mouse tumour. Thus five fraction doses given over a nine day period (5F/9d) gives much better tumour control than five daily fractions (5F/4d), at total doses which cause similar skin reactions in both schedules. Such multifraction studies can provide even more 'relevant' information for the better understanding of fractionated radiotherapy, so long as the difference between mice and men is not forgotten.

Metastases

Until there is strong evidence to disprove it, the assumption can be made that metastatic deposits have a similar radiosensitivity to the primary tumour from which they originated. In terms of *radioresponsiveness* this would mean that a given volume of metastatic tumour should respond to the same extent as the same volume of primary tumour. The clinical situation very rarely allows such a comparison to be made — primary and secondaries are not often the same size at the same time. Patients with multiple metastases which require treatment provide a commoner opportunity for objective comparison and some of the fractionation problems to be discussed in Chapter 13 can be answered just as well, or better, by using such clinical material, as by the use of animal tumours.

At the cellular level, the response of metastasising tumours to radiation has quite often been studied with animal models. The two assay methods for leukaemic cells described in Chapter 6 both involved a 'metastatic situation'. In both cases, after irradiation of donor mice bearing a primary tumour, a known number of the cells were injected into recipient mice where they would circulate and 'seed' in various organs. In one case the spleen is examined for colonies, in the other infiltration of the liver is the cause of death. (There is yet another technique in which tumour nodules grow in the lungs.) This is obviously an indirect method of studying the effect of radiation on metastases since the recipient animals never have a 'primary'; they merely provide the test bed for the viability of the irradiated cells. But although these are irradiated before they 'metastasise' they can be considered to have spread from a primary in the same animal. Their radiosensitivity certainly shows no change whichever method is used for their assay.

The best animal model for the study of distant metastases is the R1B5 fibrosarcoma since after transplantation the primary tumour grows subcutaneously, with a capsule and stroma like a clinical tumour. It is therefore possible to perform an apparently radical excision; local recurrence is uncommon after this procedure. Such operations are quite commonly followed by pulmonary and other distant metastases, but the incidence of these metastases is significantly reduced if the animals are irradiated pre-operatively (Table 10.3). The use of *pre-operative radiotherapy* before radical surgery is often intended to reduce the incidence of local recurrence since this can more certainly be attributed to malignant cells newly 'spilt' at the time of operation. (The question of wound healing after pre-operative radiotherapy was discussed in Ch. 8). In most patients it is not easy to tell whether distant metastasis was not already

Table 10.3 Death from distant metastases after excision of R1B5 sarcoma

Pre-operative dose	No. of animals	No. of deaths	% deaths
Nil	43	39	91
2000 rads	14	7	50
4000 rads	17	4	24

(from Thomlinson, 1966)

present at the time of surgery, but the results shown in Table 10.3 show an obvious reduction in metastasis following pre-operative radiotherapy.

It must be said, however, that the best information on the direct response of human metastases remains that obtained from the irradiation of lung metastases clinically. This has shown that the kinetics of the human tumours remains unchanged, after the period of growth restraint, from that which existed before radiotherapy.

Kinetics of irradiated tumours

Throughout this book emphasis has been placed upon the cellular basis of radiobiology. Tumours consist of tissues, more or less organized, with a stroma of connective and vascular tissue. When it comes to improving the therapeutic ratio between tumour and normal tissue response, however, it is the total tumour cell mass that is the critical target; limited by the reaction of the essential normal cell populations and tissues. Figure 10.7 follows the peripheral blood count of a patient with chronic myeloid leukaemia who received radiotherapy to the spleen at the times marked. On each occasion there was an immediate fall in the white cell count followed by a gradual rise to the starting

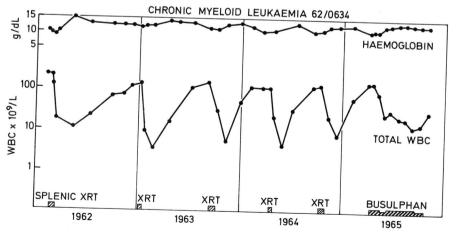

Fig. 10.7 Blood count after irradiation for chronic myeloid leukaemia.

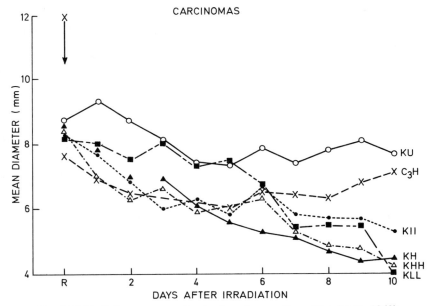

Fig. 10.8 Radiation response of animal carcinomas (from Denekamp, 1969).

level when treatment had to be repeated (until Busulphan therapy was instituted). This illustrates the concept of the 'total cell mass' in a very obvious way; not often possible to record with solid tumours. The question arises why some tumours respond in this almost predictable way and some do not.

Figures 10.8 and 10.9 show growth curves for a selection of animal carcinomas and sarcomas after 1500 rads single dose of radiation. Both figures have a linear vertical scale of tumour diameter. Amongst the carcinomas in Figure 10.8 is the C_3H mammary tumour already described in this chapter, and the R1B5 fibrosarcoma is also included in the sarcomas in Figure 10.9. These two figures show a wide scatter of results but it can be concluded that the carcinomas regressed immediately after irradiation whilst the sarcomas all continued to grow for several days before slowing down to a plateau or shrinking. Studies on the kinetics of all the tumours represented in both figures lead to the conclusion that carcinomas have a higher *cell loss factor* than sarcomas and that radiotherapy accentuates this. For the C_3H carcinoma the cell loss factor starts at 70 per cent and rises to 76 per cent; for the R1B5 sarcoma there was no cell loss at all before irradiation but a factor of 64 per cent was measured afterwards (in Figure 10.9, SSB_1 is a sub-line of $R1B_5$ and in this sarcoma the cell loss rose from 25 per cent to 68 per cent). The consequences of rapid shrinkage, due to cell loss, on tumour reoxygenation, were mentioned earlier.

This shrinkage of an irradiated tumour may be due to any one of the three processes of Apoptosis, Autolysis and Phagocytosis. Apoptosis just means the shedding of cells, i.e. cell loss in the simple sense. The process of autolysis depends on a good blood supply for diffusion; if the blood supply is poor then 'coagulative necrosis' will occur. Finally, phagocytosis requires a new growth of

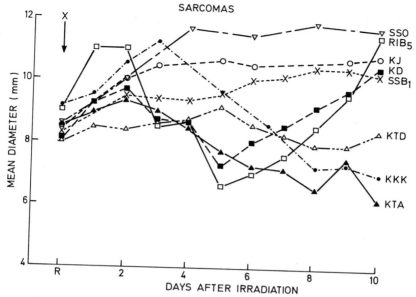

Fig. 10.9 Radiation response of animal sarcomas (from Denekamp, 1969).

capillaries and connective tissue leading to 'organisation' in the irradiated tumour volume.

Returning to Table 10.1 at the beginning of this chapter, it might be concluded that the reason for the greater response of squamous cell carcinomata to radiotherapy lies in the high cell loss factor of such a histological type. This serves to explain the radiation responsiveness of reticuloses and embryonal tumours too. Adenocarcinomata and sarcomata may be less responsive because of their lower values for cell loss factor. At the cellular level all these histological types show a close similarity in the D_o values of their survival curves although the extent of the shoulder portion has yet to be determined for many human tumours *in vivo*.

The various animal tumours provide examples of the many possible explanations for the variation in radiation response between different types of human tumours and between different patients with the same type of tumour. Animal tumour data are more accurate and more easily obtained. Many of the data can be obtained from human material, however, if radiotherapists use their clinical opportunities in a systematic manner to study the principles discussed in this chapter.

REFERENCES

Denekamp, J. (1969) in Time and dose relationships in radiation biology as applied to radiotherapy, p. 145. New York: *Brookhaven National Laboratory.*
Deschner, E. E. & Gray, L. H. (1959) Influence of oxygen tension on X-ray-induced chromosomal damage in Ehrlich ascites tumor cells irradiated in vitro and in vivo. *Radiation Research,* **11,** 115-146.

Hermens, A. F. & Barendsen, G. W. (1969) Changes of cell proliferation characteristics in a rat rhabdomyosarcoma before and after X-irradiation. *European Journal of Cancer,* **5,** 173-189.

Howes, A. E. (1969) An estimation of changes in the proportions and absolute numbers of hypoxic cells after irradiation of transplanted C_3H mouse mammary tumours. *British Journal of Radiology,* **42,** 441-447.

Malaise, E. P., Chavaudra, N. & Tubiana, M. (1973) The relationship between growth, labelling index and histological type of human solid tumours. *European Journal of Cancer,* **9,** 305-312.

Thomlinson, R. H. (1961) The oxygen effect in mammals in Fundamental aspects of radiosensitivity. *Brookhaven Symposia in Biology No. 14.* 204-216.

Thomlinson, R. H. (1966) Personal communication.

Thomlinson, R. H. (1969) Reoxygenation as a function of tumor size and histopathological type in *Time and dose relationships in radiation biology as applied to radiotherapy,* p. 243. New York: Brookhaven National Laboratory.

Van Putten, L. M. (1969) in *Time and dose relationsips in radiation biology as applied to radiotherapy,* p. 250. New York: Brookhaven National Laboratory.

11. Late Effects on Normal Tissues

The immediate cellular effects of radiation on living tissues have already been described in Chapter 8. They are normally self-limiting and are repaired within six weeks of the completion of a course of treatment. Late radiation changes in normal tissues and organs may be recognized usually no earlier than three months after irradiation, but in respect of other effects such as leukaemogenesis and the induction of cancer, (described in Ch. 12), a latent period of 8 to 10 years or even longer may be observed. These late effects are the manifestation of progressive degenerative processes induced by the radiation and their incidence increases with time, to a maximum and then declines. Occasionally late effects in an organ may be seen directly following a severe immediate reaction.

There is an indirect relationship between the degree of early reaction in a tissue and the probability of developing a late reaction, but the pathogenesis of early and late reactions is different. The immediate radiation reaction depends predominantly on the number of parenchymal cells killed by radiation, whereas the development of late reactions is also influenced by two other processes associated with progressive secondary damage (Table 11.1). These

Table 11.1 Pathogenesis of early and late radiation effects

Early effects		
Differentiated Cell	Cell depletion⟶	HYPOPLASIA
Vascular endothelium	Increased permeability⟶	OEDEMA
Late effects		
Differentiated Cell	Cell depletion⟶	ATROPHY
Vascular endothelium	Increased permeability⟶	OEDEMA
		FIBROSIS
	Endarteritis ⟶	ISCHAEMIA

changes are associated with ischaemia and fibrosis produced in the tissue. The ischaemia results from damage to the endothelial cells and walls of the blood vessels. Fibrosis may well also be directly related to the degree of cell killing in the vascular endothelium, resulting in protein fractions extravasating into the interstitial tissues. It must be noted, however, that the pathogenesis of radiation fibrosis is not properly understood. The actual late radiation effect will depend on the relative damage to the parenchymal cells and to the vascular endothelium. This will differ from tissue to tissue and may also be influenced by the quality of radiation and the method of dose fractionation. The interpretation of the differences seen in late reactions in normal tissues will depend on the relative extent of damage to the specific tissue cells and to their supporting vasculo-connective tissue.

Effects on vascular tissue

Some of the gross effects of radiation on vascular structures have been described in Chapters 3 and 8. It has been noted that the larger vessels, the arteries and large veins, show less radiation damage than the arterioles and the capillaries. In the fine blood vessels, death of the endothelial cells leads to irregular proliferation of surviving cells and intimal thickening. Damaged endothelial cells under the intima may absorb lipid material and become large 'foam cells' which are often a feature of radiation-induced vascular damage. Damaged cells in the intima may later die and give rise to vacuolization of the intima. In fact, all layers of the vessel wall will be damaged by the radiation and in due course all the features of vascular sclerosis will develop, at times with evidence of intra-vascular thrombosis. The radiation effect on cells is a random process and the irregularity of cell killing along the length of the arterioles and venules is later seen as a series of dilatations and 'sausage-shaped' segments. In the capillaries the features are those of telangiectasia the severity of which depends upon the dose delivered.

In terms of cellular response it has been demonstrated (Fig. 11.1) by using an *in vivo* system that the endothelial cell lining small blood vessels of rats has a D_o value of 170 rads and a D_q value of 340 rads, (Reinhold & Buisman, 1973). It will be remembered that the turnover of these cells is rather slow with a cycle

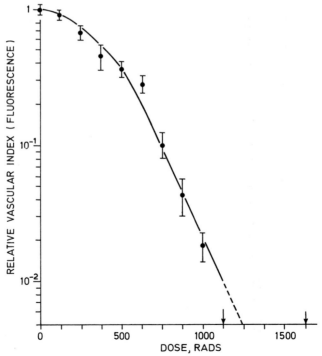

Fig. 11.1 A 'survival curve' for capillary endothelial cells, derived *in vivo* by a microfluorometry of an intravenously injected tracer after single doses of radiation. (300 kV h.v.l. 3mm Cu). (From Reinhold and Buisman. (1973)

time of 50 hours and so they will not, from the point of view of the immediate reaction, appear to be highly radioresponsive. In the development of late effects, however, their response is of prime significance.

It should be pointed out, too, that intercurrent diseases that are accompanied by generalized endarteritis, such as diabetes mellitus and hypertension from any cause, will aggravate the effect of radiation damage on the vasculature and so lower the radiation tolerance of vascular tissue.

Radioresponsiveness will depend in the final analysis upon a combination of the particular vascular architecture of the tissue and the complex interaction of the mechanisms of cell death, recovery and repopulation (and reoxygenation in the case of tumours) which were discussed as separate phenomena in earlier chapters. The results of such interactions will be discussed now for various organs of importance in radiotherapy.

Bone and cartilage

In young animals, bone is formed mainly from epiphyseal cartilage. Chondroblasts divide, arrange themselves in columns, and form a matrix of collagen and mucopolysaccharides in which hydroxyapatite is deposited. This primary bone structure is subsequently remoulded by osteoblasts which deposit and by osteoclasts which remove bone. The latter reactions proceed at a reduced rate throughout adult life.

Cell renewal in chondroblasts is easily affected by irradiation. The radiobiological characteristics of chondroblasts have been assayed by clonal techniques and a D_0 of 165 rads and an extrapolation number n of 6 were found. Since one column in the epiphyseal plate consists of about fifteen chondroblasts, at least one cell in each column must remain intact after exposure to less than 800 rads for growth to be merely delayed without being permanently impaired. After exposure to higher doses (up to about 2 k rads) the architecture of the epiphyseal pleate is disturbed, and growth is arrested for several weeks.

Non-growing portions of bone are relatively radioresistant but if large doses of radiation are given to a growing bone, most cartilage cells degenerate, osteoblasts are destroyed, and an acellular, avascular bone-like substance may be formed from the cartilage. This will gradually be resorbed and replaced by bone to form a solid plate which is incapable of growth. The result will be a shortened bone. If some of the resting cartilage cells survive the radiation, they may migrate distally to form a new functioning epiphyseal line.

Irradiation of formed bone may result in a derangement of the synchronization of resorption and new bone deposition. This may produce either excessive absorption or an overgrowth of bone. In irradiated bone, fractures are common due either to structural damage to the bone matrix or to defective bone mineralization.

Central nervous system

The central nervous system was considered at one time to consist of tissues of high radio-resistance. It is now recognized that this false interpretation resulted from the commonly late manifestation of radiation injury which is related to the slow turnover of cells in nervous tissue.

The effects of very high doses, of the order of 10 000 rads single exposure have been described in Chapter 9. This level of exposure leads rapidly to acute meningo-encephalopathy. It is characterised by oedema of all intra-cranial structures and increased production of cerebro-spinal fluid, increased pressure and death in most cases is due to coning of the mid-brain. There is gross vascular damage, increased permeability of the blood-brain barrier and diffuse neuronal destruction.

Exposure to non-lethal doses may produce a sub-acute meningo-encephalopathy (Lampert & Davies, 1964) that is a rapidly demyelinating process from which, however, complete clinical recovery is possible.

When the brain is irradiated to the high levels of dosage used in cancer therapy there is the risk of late *brain necrosis*. (Boden, 1950). Chronic radiation encephalopathy usually presents 3 to 24 months after high-dose irradiation; the higher the dose the shorter the latent period. The injury is strikingly selective of the white matter indicated by widespread demyelinisation. The cerebral cortex is least affected while the brain stem is particularly vulnerable. The blood vessels are also normally seen to be damaged with all degrees of degenerative change previously described and may include complete thrombosis, which will have grave clinical consequences in the brain.

Similar changes may be found in the spinal cord. Sub-acute radiation myelopathy is usually a transient condition which occurs 2 to 4 months after irradiation, and may persist for several months. (Jones, 1964). It is considered to be caused by the inhibition of myelinisation due to loss of oligodendroglia and restitution occurs with time.

Chronic *radiation myelopathy* denotes severe and irreversible damage to the spinal cord (Phillips & Buschke, 1969). The symptoms and signs are of partial and eventually complete transection of the cord. The white fibre tracts are greatly reduced, the grey columns standing out as pale areas, but often containing petechiae. The loss of tissue may be so great that the spinal cord almost disappears.

At the cellular level these changes can be attributed to a greater sensitivity of oligodendrocytes in white matter than those in grey matter, with consequent demyelination. Vascular damage may lead eventually to ischaemic changes, but the acute response produces oedema due to increased vascular permeability.

Depending upon the dose level there are two modes of radiation injury to the central nervous system. One mode is vascular which predominates with single exposure of between 2000–3000 rads. The other affects white matter through damage to the supporting glia and is most clearly seen with single exposures of 3000–4000 rads. The two modes of damage can be explained in cellular terms and a cell-survival model may be assumed to explain the incidence of radiation myelitis as a function of dose. A survival curve for neuronal cells has been derived that is consistent with a D_o value of 130 rads and an extrapolation number of 2, although it cannot be regarded as conclusive.

The volume of brain or spinal cord irradiated is an important factor in the tolerance dose of radiation. Boden (1948) suggested that, for orthovoltage radiation, the threshold dose was 4500 rads in 17 days for fields less than 50

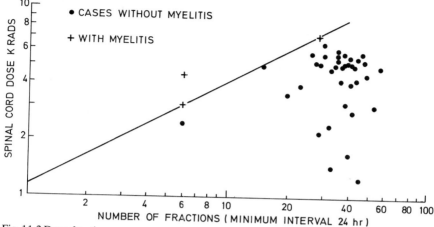

Fig. 11.2 Dose-fraction response curve for irradiation myelitis of the spinal cord in the thoracic region. The line extrapolates from the lowest dose points after which myelitis was observed. (From Phillips and Buschke (1969).

cm^2 and should be reduced to 3500 rads when the fields were more than 100 cm^2. The tolerance doses in respect of thoracic spine are illustrated in Figure 11.2.

Thoracic organs

The lungs, oesophagus and heart will be considered under this heading because irradiation of the chest will usually involve all three of these organs. The effects of thoracic irradiation can be contrasted with the lethal responses to total body irradiation which was described in Chapter 9, where a well-defined sequence of CNS, gut and bone marrow syndromes was shown to have a cellular basis in which the response to progressively higher dosage can be related to the shortening time scales associated with the three modes of death. Similarly, for thoracic irradiation, an early 'starvation death' occurs after higher dose levels which produce principally oesophageal damage, while a later mortality is seen in experimental animals to occur from lung fibrosis after lower doses. For thoracic irradiation, then, there are 'oesophageal' and 'pulmonary' syndromes and these can also be shown to have a cellular basis.

Lung

The adult lung is a stable tissue with a very slow cellular proliferation in its differentiated cells and in the endothelial cells. This contrasts with the oesophagus which has a very rapid cell renewal system. Consequently the effects of radiation appear early in the oesophagus whereas the effects on the lung are seen much later.

Radiation pneumonitis follows damage to either or both the cell populations found in the alveolar septa of the lungs: these are endothelial and alveolar cells. The response of capillary endothelial cells has already been mentioned and their damage may be the major factor in the development of pneumonitis. This may not be the only interpretation, however. The alveolar cells include a type which are vacuolated and secrete lung surfactant, a liquid material which, by reducing the surface tension of the fluid layer lining the alveoli, maintains their stability. Hydrostatic and osmotic pressures are balanced so as to prevent collapse of the alveoli upon expiration. Thus radiation pneumonitis may be just as much the consequence of dose-dependent loss of these alveolar cells as the result of capillary endothelial cell loss. In either or both cases there will be damage to the lung stroma characterised by oedema followed by hyalinization and fibrosis of the alveolar walls (Phillips & Margolis, 1972). Resulting impairment of ventilatory and diffusion capacities of the lung may be significant in the

Table 11.2 The development of radiation pneumonitis (from Van den Brenk, 1971)

Phase	Sequence of events	Time span
Exudative	Cell damage———→Inflammatory exudates	0–40 days
Pneumonitis	Desquamation, Consolidation, Organisation	20–60 days
Fibrosis	Fibrosis, Devascularisation	60–200 days
Secondary changes	Calcification, Metaplasia, Neoplasia	> 200 days

long-term effects of radiation (see Table 11.2). Haemorrhage into the lung may occur, also, as a result of radiation changes in the blood vessels.

Differences between the radiation response of lung and oesophagus are also evident in terms of dose response data from rats and mice, as well as from clinical data (Fig. 11.3). Such comparisons permit the calculation of a dose of 900 rets (NSD) for a 5 per cent incidence of clinical pneumonitis and 1040 rets (NSD) for a 50 per cent incidence.

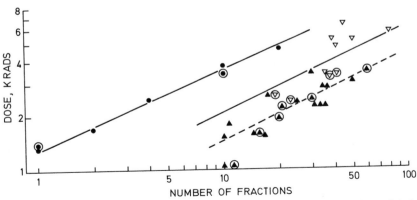

Fig. 11.3 Dose-fraction response curve for radiation pneumonitis in the mouse (closed circles) and human subjects (triangles). The mouse data is measured by the LD50/160. The open triangles indicate clinical cases of pneumonitis, closed triangles cases without pneumonitis. The circles indicate the addition of Actinomycin D, which reduces lung tolerance in patients. (From Phillips and Margolis (1972)

In rodents death due to pneumonitis occurs late; between 40 to 180 days after irradiation. The process is slow and there is also a slow repair of sub-lethal damage when fractionated radiation is used. This slow repair appears to supplant the effect of cellular proliferation in this organ.

Heart

Injury to the heart following thoracic irradiation is most often manifested as pericarditis with fibrosis which is self-limited. The pericardial thickening is due to dense collagen which replaces the pericardial adipose tissue and organised fibrinous exudate. A pericardial effusion is normally present. Clinically these late cardiac lesions may present as an acute exudative pericarditis, or as a chronic constrictive *pericarditis.*

Damage to the myocardium may also be seen and usually radiation changes in the fine vessels precede the onset of fibrosis and the myocardial degeneration is secondary to ischaemia.

Stewart and Fajardo (1973) correlated the probability of cardiac damage with radiation exposure. When only a small segment of the heart is irradiated, such as in breast cancer treatment, a dose of 1850 rets (NSD) will result in mild carditis in about 5 per cent. When at least half the heart is included in the X-ray field the dose is reduced to 1500 rets (NSD) for the same incidence of carditis.

Oesophagus

The acute effects of radiation on the oesophagus is seen within a week of a single dose of irradiation and within ten days the oesophagitis is usually at its height. The degree of desquamation of the epithelium will be dependent on the size of the dose delivered.

Sub-acute and chronic radiation oesophagitis is characterized by persistent ulceration commonly associated with stenosis of the lumen. As in the intestine these ulcers tend to be penetrating and perforations do occur. In addition to loss of epithelium, a striking feature of the reaction is submucosal oedema. With very high radiation doses damage to the muscle coat is also seen in the form of cellular degeneration and hyalinization.

In contrast to damage to the lungs, much higher dose levels are necessary for severe late oesophageal complications. Phillips and Margolis (1972) have calculated from clinical data that radiation doses equivalent to about 1850 rets (NSD) are likely to produce late oesophegeal effects in 5 per cent of patients. They suggest that should the radiation dose be increased to 2000 rets (NSD) it is probable that 50 per cent of patients would suffer serious late damage to the oesophagus.

In mice oesophageal damage is seen one to two weeks after a large single dose of X-rays and death occurs 10 to 20 days after irradiation. The LD50 for these mice is 2500 rads delivered to the thorax in a single exposure.

Intestine

The immediate reaction of the intestine to radiation has already been described in relation to the acute radiation syndromes in Chapter 9.

Severe late effects of radiation on the intestine may be clinically evident as early as three months following completion of a course of treatment (Roswit *et al.*, 1972). Sub-acute enteritis is characterized by gross epithelial damage, submucosal oedema and some degree of vascular degeneration. Clinically it is manifest by persistent or recurrent diarrhoea, occasionally associated with bouts of pseudo-obstruction. Chronic radiation enteropathy may be characterized by two different pathological forms, ulcerative and fibrotic, depending on which feature predominates. In no case is one or the other form seen exclusively and the clinical features tend to be common. Recurrent diarrhoea, abdominal pain and wasting are found and malabsorption may be present. In patients where the epithelial damage is most obvious the most striking feature is gross atrophy of the epithelial cells with loss of the villi in the small intestine. Ulceration is always a feature, at times in the form of tiny superficial lesions, but in some patients there may be the formation of penetrating ulcers. These ulcers characterize radiation lesions and they may lead to haemorrhage, perforation of fistula formation. Often they may be large necrotic ulcers with heaped-up margins which may be mistaken for malignant change.

In about 10 per cent of patients who have radiation enteropathy fibrosis of the bowel wall, particularly in the submucosa, is the significant feature. In these patients intestinal stenosis results and may appear as either annular or tubular areas of narrowing of the lumen.

The acute radiation effects on the intestine are readily explained by the cellular depletion of the epithelium. There is an immediate decrease in the numbers of dividing cells and early death of the cells in the crypts of Lieberkuhn. If the radiation dose is high enough, say 1000 rads single exposure, there is rapid cell lost in the intestinal crypts and the intestinal villi become short and blunted. Absorption defects and bacterial invasion of the bowel wall may be detected, associated with excessive loss of fluid and electrolytes. Late effects are normally seen in patients who have had some degree of acute or sub-acute enteropathy and the lesion is a combination of epithelial damage associated with vascular deficiency and fibrosis. Radiation changes in blood vessels in the bowel wall are common features of late radiation enteropathy.

In relation to the cellular response of the intestinal (jejunum) epithelium, Withers & Elkind (1969) have shown in the mouse a sensitivity similar to other cells with a D_o value of 115 rads and a high capacity for repair of sublethal damage ($D_q = 430$ rads) following large single doses of X-irradiation (Fig. 11.4). After exposure to fast neutrons it is interesting that the D_o value 100 rads is very similar, but after equivalent dosage the D_q falls to 200 rads indicating a much reduced capacity to shed sub-lethal damage. (see ch. 6 and figure 6.4 for discussion of the shape of this sort of curve.)

Kidney

Late radiation injury to the kidney is commonly referred to as radiation 'nephritis', but it may be manifest in several different ways and is really better described as radiation nephropathy. Hypertension is often a feature and radia-

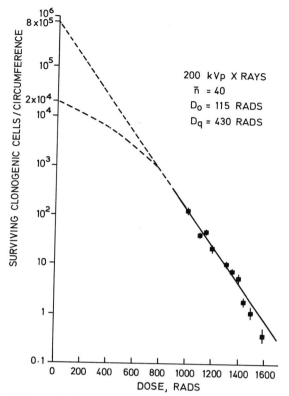

200 kVp X RAYS

$\bar{n}$ = 40

D_0 = 115 RADS

D_q = 430 RADS

Fig. 11.4 Cell survival curve for intestinal epithelial cells derived from *in vivo* irradiation of the mouse jejunum.
(From Withers, H. R., & Elkind, M. M. (1969).

tion injury to the kidney is particularly likely to produce hypertension (Luxton & Kunkler, 1964).

The kidney is an organ with many highly specialized functions and its cell systems all have very slow turnover rates. As a result radiation injury is usually not seen until some months after exposure. However, the incidence of renal damage and its time of presentation after irradiation has been shown in mice to depend on the dose of irradiation (Fig. 11.5).

Following single doses of X-rays below 1000 rads few changes can be seen in the kidney and it is usually only after higher doses that impairment in renal function is found. Doses in excess of 1000 rads to *both* kidneys are likely to lead to renal failure in a high proportion of patients.

The clinico-pathological types of renal damage may be classified as 'acute' and 'chronic' radiation nephritis.

Acute radiation nephritis or progressive glomerular sclerosis has a latent period of six to twelve months and is associated with proteinuria, anaemia and hypertension. Patients rapidly become seriously ill after the onset of symptoms and about 30 per cent will die in the acute illness. All who recover will have chronic radiation nephritis, but in about half the blood pressure may be

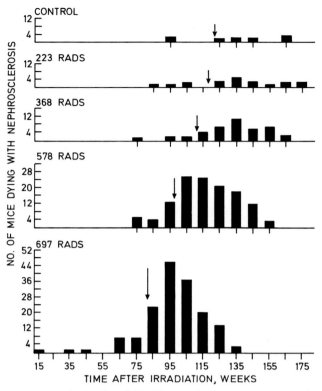

Fig. 11.5 Age distribution of death from radiation nephropathy in mice. (From van Cleave (1968)

brought under control. The kidneys in patients with acute radiation nephritis are normal in size and usually little is to be seen microscopically. Microscopy may show damage to the glomeruli and tubules with fibrinoid necrosis in the arterioles often a prominent feature.

Chronic radiation nephritis may take the form of either nephrosclerosis or sclerosing nephrosis.

Radiation-induced nephrosclerosis may follow the acute syndrome, or present after a latent period of eighteen months or more. The onset is usually insidious with mild proteinuria, anaemia and hypertension and the syndrome may be compatible with normal life for many years. Some patients may eventually die of renal failure and perhaps 25 per cent may develop 'malignant' hypertension. In patients in whom only one kidney is damaged, nephrectomy may reverse the progress of the disease. In these patients the kidneys typically are small and scarred with thickened capsules. Microscopically there is a glomerular hyalinization tubular atrophy with gross fibrosis and usually marked fibrinoid degeneration in the vessels.

Sclerosing nephrosis is the mildest form of radiation damage to the kidney. Mild proteinuria may be found, but may be intermittent. Renal function tests are often normal, although the blood urea may rise for example in times of stress. Hypertension may be found and is usually mild, although some patients may develop malignant hypertension. The kidneys are apparently normal and it is only by special techniques that some degree of interstitial fibrosis and glomerular atrophy may be seen.

It has not been established whether the hypertension following irradiation of the kidney is a primary or secondary effect. Hypertension after irradiation has been shown to be associated with high levels of angiotension, but this may be the result of progressive renal ischaemia due to radiation changes in arterioles. This would seem a more probable explanation than any primary effect on the cells of the juxta-glomerular apparatus.

Liver

The liver has a very low mitotic index and so most hepatocytes will be in the relatively resistant, inter-mitotic period at the time of irradiation. Because of the extremely slow turnover of liver cells, any radiation damage is usually manifest late. Human liver cells, however, have been cultured and the D_0 value *in vitro* estimated to be 119 rads, demonstrating the intrinsic radio-sensitivity similar to most other mammalian cells.

Acute radiation hepatitis may follow high radiation doses to the liver. In these cases the effect is primarily seen in the vascular tissue, there being marked congestion, hyperaemia and occasional haemorrhage. Some degree of hepatic cell atrophy in the centrilobular regions may be evident.

Chronic radiation hepatitis is most often found in part of the liver which has been in the field of high dose irradiation. The pathology seems to suggest that the predominant lesion is again in the vasculo-connective tissues. The arteries and veins showed evidence of radiation damage, associated with interstitial fibrosis and thickening of Glisson's capsule. There is also atrophy of the liver cells, but often nodules of regenerating liver cells may be seen in the atrophic and fibrosed part of the liver.

Special mention has to be made of the effects of radioactive metal colloids on the liver after parental administration. These substances are concentrated in the Küpffer cells of the liver and produce serious progressive changes. Thorium was the most widely used substance of this type for it provided excellent contrast for radiography. Thorium, which is an alpha-particle emitter, is retained in the reticulo-endothelial cells and an interstitial sclerosis supervenes due to continuous irradiation damage. The effect of Thorotrast, as the commercial contrast medium was called, often progressed to the development of cancers in the liver. Carcinomas have been reported, but more commonly the hepatic tumour in these cases is found to be a haemangio-sarcoma.

Thyroid gland

Morphology and function of the normal thyroid appear to be relatively resistant to direct effects of irradiation, whereas the hyperactive thyroid responds more readily and stimulated cell renewal in the thyroid is as radiosensitive as it is in other organs. High doses of radiation can, however, damage even the normal thyroid permanently and eventually cause hypothyroidism, characterized by a flat epithelium, few follicles, and an increase in connective tissue. Whole-body irradiation appears to modify the normal thyroid more conspicuously than local exposure; the thyroid then shows a histological structure which suggests important functional changes. Most likely, however, these functional changes following whole-body exposure result from abscopal mechanisms mediated by the pituitary.

The radiosensitivity of thyroid tissue has been determined using rats subject to goitrogenic stimulus. The normal rat thyroid shows little change after irradiation judged by weight, cell counts or DNA and RNA content, because proliferation does not occur in the adult gland (i.e. there is an out-of-cycle (G_o) population). The effects of radiation on rat thyroid proliferative capacity can be measured if the animals are given methyl-thiouracil to promote cell multiplication. With this goitrogenic stimulus dose-response curves have been obtained with respect to gland weight and these can be used to derive cell survival curves for single doses of neutron and gamma irradiation (Fig. 11.6). The parameters of the gamma ray curves are $D_o = 405$ rads, $N = 2.8$; for fast neutrons $D_o = 310$ rads, $N = 1.3$. (Typically, RBE values fall with increasing dose from 3.1 to 1.7).

Lens and cataract formation

The most prominent radiation effect on the eye is the development of lens opacities (Upton, 1969). Much of the experimental studies on radiation cataractogenesis has used the murine lens as the model system, but it should be noted that the murine lens is particularly sensitive in this respect. The sensitivity of the lens is very much related to species and in larger mammals, including man, the lens is much more resistant to radiation damage. The sensitivity of the lens is also age dependent, the older the animal the greater the radiation effect and the shorter the latent period after irradiation.

Low doses of X-rays produce minimal lens change in mice, months or even years after exposure, but severe alterations are produced by moderate to large doses within 6 months to a year and neutrons are particularly effective in producing cataracts. Normally, doses greater than 500 rad of X-rays are required to produce clinically significant cataracts, but lens opacities have been reported in man after as low as 200 rad of mixed γ and neutron irradiation. On the other hand, the protracted radiation from a radon seed implant in the eye may not lead to cataract below a dosage of 2000 rads to the lens. In most cases, lens opacities develop after a latent period between exposure and the appearance of the cataract which is an inverse function of dose. While in humans the average latent period is 2 to 3 years the time of onset may range from 6 months to many years, the interval being related to the dose and overall time of treatment. The severity of lens opacities in mice in relation to dose and time after exposure is illustrated in Fig. 11.7.

The anterior epithelial cells differentiate into the fibres of the lens and it is this population which provides the cellular basis for the formation of radiation

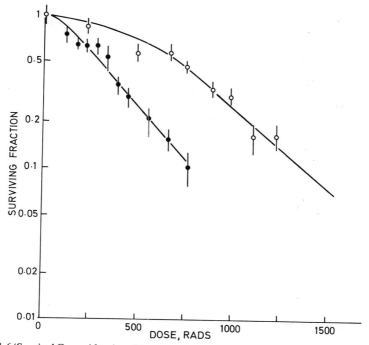

Fig. 11.6 'Survival Curves' for thyroid cells derived *in vivo* by goitrogenic challenges following single doses of γ-rays (open circles) and 14.7 MeV fast neutrons (closed circles).

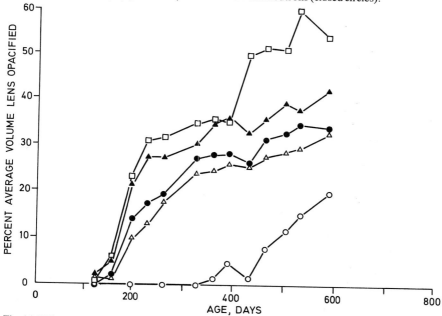

Fig. 11.7 The severity of radiation cataracts formed in mice after irradiation with different doses of X-rays.
O controls, △50 rads, ●100 rads, △200 rads, □ 400 rads.
(From Upton (1969).

cataracts. Following irradiation there is a decrease in the mitotic activity of the germinative cells and abortive attempts to elaborate into normal lens fibres are seen. As a result there is accumulation of abnormal cells and debris which becomes visible when they reach the posterior pole. With higher doses, all the structures within the capsule may eventually be involved and the lens may become completely opaque in a progressive deterioration which may continue over 10 years or more after irradiation.

Neutrons have been reported to be particularly effective in producing cataracts and very high RBE values have been reported from experiments on mice. However, since the murine lens is so sensitive to this effect, extremely low doses of radiation have been used and therefore very high RBE values are obtained because of the difference in the shape of the initial slopes of the survival curves of X-rays and neutrons. In man it seems likely that the RBE for cataract formation may be of the same order as that for other cellular effects, but careful clinical observations will have to be made.

Testis

In the male gonad there is a stem cell population and progeny which are customarily categorized into fourteen spermatogonial stages A_o——Y A_4 and B-spermatocytes, spermatids and finally the mature spermatozoa. Spermatogonia are the most sensitive cells and are killed after exposure of the testes to relatively low doses of radiation and die in early prophase or later metaphase. As the supply of germ cells derived from spermatogonia becomes exhausted, the testes are progressively depleted of these cells until at 2 to 4 weeks after exposure mature sperm cells have disappeared. If the dose has not been excessive, regeneration from type A spermatogonia spared from radiation death begins. Fig. 11.5 is a schematic diagram showing this course of events in the mouse, as it affects sperm production for ejaculation. The time course of this phenomenon over a five-year period has been described for a man who had received an accidental dose of total body irradiation (Oakes & Lushbaugh, 1952) (Fig. 11.8). The sperm count reached its lowest level six months after irradiation and then recovered a normal level at 2 to 3 years. In contrast, the Leydig and Sertoli cells of the interstitial tissues are relatively radio-resistant; testes atrophied because of radiation damage appear to contain more interstitial cells than germ cells.

Histological examination of mouse testes after irradiation shows that regeneration of spermatogenic epithelium occurs in discrete foci. On this basis a single dose 'cell' survival curve has been derived and this has a D_o value of 180 rads. Fractionation studies show that recovery from sub-lethal damage also occurs in this tissue and the value of D_2-D_1 is 270 rads after an interval of 4 hours.

In the human male a dose of 200 rads is likely to produce temporary sterility for about 12 months. A dose of 500 rads will produce permanent sterility in most men, but their libido and potency is normally retained as the interstitial cells, the main source of male hormone production, would not have lost their functional integrity.

Ovaries

Early in this century it was recognized that the ovaries atrophy after irradiation and that temporary or permanent sterility may result. In order to evaluate studies of irradiated ovaries, stress must be laid upon the fact that no other organ presents such large differences in response after irradiation between species or between individuals within a species. In the developing feotus, oogonia are relatively radioresistant, but oocytes in primordial follicles are extremely radiosensitive ($D_o = 91$ rads, extrapolation number: N ; 2–3). Sensitivity diminishes as the follicles mature, and is low before ovulation. Thus, female (as well as male) germ cells are most sensitive during the last premeiotic prophase and the development into mature gametes. The crucial difference between male and female germ cells is that male germ cells represent a renewing cell population whereas female germ cells do not. The 'fertile-sterile-fertile' pattern often found in the ovaries after irradiation is not a result of regeneration from a stem-cell pool, as it is in males, but of the higher sensitivity of the intermediate follicular stages compared with that of the primitive and mature stages. Granulosa cells in the developing follicles are damaged even earlier than the oocytes, but in the mature follicles and the corpus luteum they appear more resistant.

Relatively low doses of radiation given during the second and third week after birth have drastic effects on fertility of the female mouse. This effect can

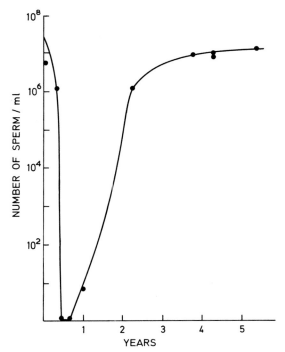

Fig. 11.8 A chart of the changes in human sperm count for 5 years after whole-body irradiation. (From Oakes and Lushbaugh (1952).

be traced to depletion of the oocyte pool by radiation-induced cell killing. Since the mouse cannot replenish losses from the oocyte pool established during embryonic life, sterility ensues when the supply of oocytes surviving radiation is exhausted.

The minimum dose to produce permanent sterility in women is not known, but it is known that the dose-effect will depend on the age of the woman (Peck et al., 1940). A dose of 200 rads will produce temporary amenorrhoea in a young woman, and 500 rads will result in permanent amenorrhoea in about 30 per cent of young women in the 30 to 35 age group. In pre-menopausal women aged 35 to 40, a single exposure of 500 rads will induce the menopause permanently in 80 per cent.

REFERENCES

Boden, G. (1950) Radiation myelitis of the brain stem. *Journal of the Faculty of Radiologists*, **2**, 79–94.

Jones, A. (1964) Transient radiation myelopathy. *British Journal of Radiology*, **37**, 727–744.

Lampert, P. W. & Davis, R. L. (1964) Delayed effects of radiation on the human central nervous system. *Neurology*, (Minneap) **14**, 912–917.

Luxton, R. W. & Kunkler, P. B. (1964) Radiation nephritis. Acta Radiologica (Stockh.), **2**, 169–178.

Oakes, W. R. & Lushbaugh, C. C. (1952) Course of testicular injury following accidental exposure to nuclear radiations. *Radiology*, **59**, 737–743.

Peck, W. S., McGrier, J. T., Kretzschan, N. R., Brown, W. E. (1940) Castration of the female by irradiation; the result in 334 patients. *Radiology*, **34**, 176–186.

Phillips, T. L. & Buschke, F. (1969) Radiation tolerance of the thoracic spinal cord. *American Journal Roentgenology, Radium Therapy and Nuclear Medicine*, **105**, 659–664.

Phillips, T. L., & Margolis, L. W. (1972) Radiation pathology & clinical response of lung and oesophagus: *Frontiers of Radiation Therapy and Oncology*, ed. Vaeth, J. M. Karger, Basel & U.P.P. Baltimore: **6**, 254–273.

Rheinhold, H. S. & Buisman, G. H. (1973) Radiosensitivity of capillary endothelium. *British Journal of Radiology*, **46**, 54–57.

Roswit, B., Naksjt, S. J., & Reid, C. B. (1972) Severe radiation injury of the stomach, small intestine, colon and rectum. *American Journal of Roentology, Radium Therapy and Nuclear Medicine*, **114**, 460–466.

Stewart, J. R. & Fajardo, L. F. (1972) Radiation-induced heart disease. *Frontiers of Radiation Therapy and Oncology*. **6**, 274–288. Karger Basel & U.P.P. Baltimore.

Upton, A. C. (1969) Radiation cataractogenesis. *Radiology*, **80**, 610-614.

Van Cleave, C. D. (1968) *Late Somatic Effects of Ionising Radiation*. Washington: U.S.A.E.C.

Van den Brenk, H. A. S. (1971) Radiation effects on the pulmonary system in *Pathology of Radiation*. Edited by C. C. Berdjis. Baltimore: Williams & Wilkins.

Withers, H. R. & Elkind, M. M. (1969) The Response of intestine to radiation. I Radiosensitivity and fractionation response of crypt cells of mouse jejunum. *Radiation Research*, **38**, 598–601.

FURTHER READING

Radiation effects and tolerance, normal tissue. *Frontiers of Radiation Therapy and Oncology*, **6**, (1972) Edited by J. M. Vaeth. Karger. Basel & U.P.P. Baltimore.

12. Late Genetic and Somatic Effects

This chapter describes a number of radiation effects on mammalian cells that all have a profound influence on the survival of the organism. These effects are manifest by either genetic or somatic damage. A knowledge of these changes and of the incidence of their induction by irradiation determine the principles that should be observed in minimising the hazards of radiation exposure.

Genetic effects of radiation

The biochemical changes related to chromosome damage have been discussed and methods for their analysis described in Chapter 4, (Fig. 4.4). Structural changes in the chromosomes occur spontaneously and irradiation merely increases the probability of changes of the same type being produced. The changes may occur at the same time in both chromatids when they are called chromosome aberrations (Fig. 12.1). Chromatid aberrations can only be induced in cells that are irradiated at a time after chromosome replication, that is the G2 phase, after DNA synthesis, and usually chromosome aberrations will occur in cells irridiated in the G1 phase before replication of DNA. Sometimes if both chromatids are hit, chromosome type aberrations may be seen, but are

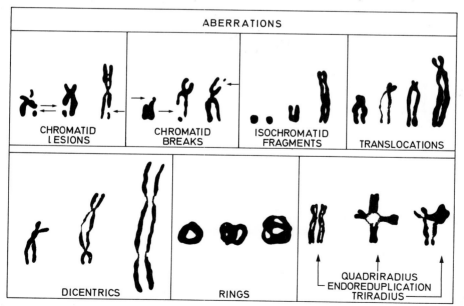

Fig. 12.1 Some examples of chromosome aberrations seen in patients during and after radiotherapy. From Amorose *et al.*, (1967).

then called iso-chromatid aberrations.

The damage to the genetic material may take the form either of exchange of parts of the DNA chain between different chromosomes or chromatids, or deletions when part of the chromosome or chromatid is broken. The exchanges may be symmetrical or assymetrical, depending on whether or not the distribution of genetic material is equal at subsequent divisions. It should be recognized that most chromosome and chromatid breaks heal in that the broken ends join up again and this is known as restitution. Rejoining readily takes place except under anoxic conditions, and the many forms of aberrations are assisted by the stickiness of chromosomes immediately after irradiation. The '*stickiness*' of chromosomes is seen in cells that are in division (particularly late prophase) at the time of irradiation, as a tendency of the chromosomes to adhere to one another in clumps. It is thought this change is due to the partial dissociation of the nucleo-protein coat that leaves molecular bonds free for linking with other chromosomes. The adhesion of chromosomes may lead to failure to complete division or at times non-dysjunction in which there is unequal distribution of genetic material between the two daughter cells.

Single break aberrations

When rejoining does not occur the two damaged ends of the chromosome or chromatid will heal resulting in what is called a terminal deletion. The portion containing the centromere is called the centric fragment and the other the acentric fragment. During division the acentric fragment will remain in the cytoplasm and its genetic material will be lost from one or both daughter cells, depending on whether the effect was in a chromatid or chromosome. It would seem that 'stickiness' is a reversible process as cells irradiated in interphase or early prophase do not show these changes when they come to enter division and it is presumed that the structure of the nucleo-protein coat has been repaired.

Double break aberrations

The name interstitial deletion is sometimes given to these aberrations where genetic material is lost between two breaks. Several types of re-arrangement may occur. Inversion of the genetic material is said to occur when the broken piece is turned around before rejoining. Ring chromosomes may occur if the broken ends rejoin and, if it contains the centromere the aberration is called a centric ring. If a fragment from one chromatid or chromosome is completely broken off by radiation and is rejoined to the end of another, the exchange is known as translocation. When a centric fragment is rejoined with another centric fragment the abnormal form is known as a dicentric chromosome that is one of the most-readily recognized forms of radiation mutations. This type of aberration may give rise to difficulties and division should the centromeres go to opposite poles and at anaphase a bridge will be formed. This bridge will have to break when the two daughter cells are formed, usually leading to some disparity in the genetic material. Other aberrations that may be seen after irradiation using the ordinary light microscopes include what are called 'gaps'. More detailed examination by electron microscopy will show a thin portion of genetic structure between the ends of the gaps. They do not therefore represent true breaks in the chromosomes or chromatids and they do not produce acentric fragments just before division. Their nature is not understood, but they do appear to be capable of restitution as their incidence falls with time after irradiation.

The incidence of chromosome aberrations depends on a number of biological factors, such as the type of cell, phase of the cell cycle, and also on the nature of the physical injury caused by the radiation. If a single hit is required the number of aberrations produced will be proportional to the dose delivered.

$$Yield = kD + C$$

Where k is a constant associated with a particular aberration and C is the spontaneous rate of that aberration—usually so small that it may be ignored.

If chromosome aberrations, such as dicentrics, require the hits of two ionising particles, their yield will be proportional to the square of the dose. $Yield = kD^2 + C$. Other changes may depend on the hits of three ionising particles and so will be related to the cube of the dose (Fig. 12.2). The measurement

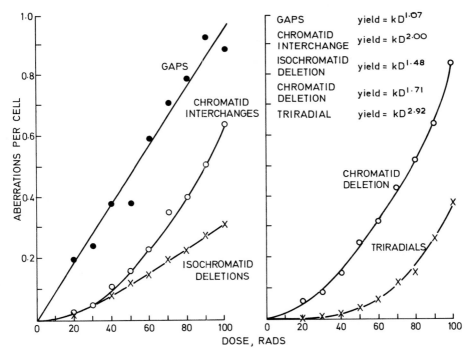

Fig. 12.2 Dose response curves from chromatid aberrations produced by radiation in *Vicia faba* root meristems. (From Revell, S. H., 1955).

of the number of dicentrics by *in vitro* lymphocyte culture is now one of the most accurate methods of determining the dose of radiation received by an exposed person. The coefficient of dicentric formation has been estimated to be 2.78×10^{-5}/cell/rad^2 and so in human lymphocyte cultures 190 rads will produce one dicentric per cell. These techniques allow an accurate assessment of radiation dose to be made retrospectively following accidental exposure.

The expression of chromosome aberrations is in gene mutation whereby some feature of the cells form or function is changed and that change is handed down in the progeny. The mutations that follow irradiation are identical to those that occur spontaneously, but their frequency after irradiation will be increased by two, three or four orders of magnitude, depending on the dose of radiation. *Chromosome mutations* are the result of either an increase or decrease in the number of genes in the nucleus. At times changes in gene structure may involve quite small bio-chemical alterations in the DNA molecules and these result in point mutations, and imply no change in the number of genes. *Point mutations* occur in simple proportion to dose and are independent of dose rate over a very wide range. These changes are also independent of any fractionation effect and of the quality of the radiation.

Effects on the embryo and fetus

Many physical, chemical and infective agents are known to carry high risks of producing damaging effects on the embryo and fetus and great care is taken to minimise such hazards. Severe effects will commonly follow the irradiation of the developing offspring. Russell (1954) has shown that the important stages of

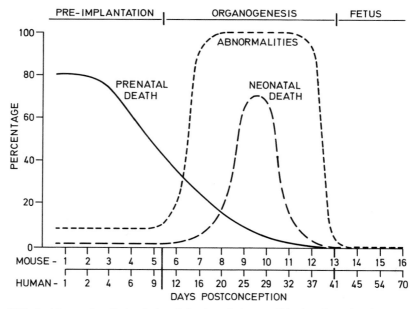

Fig. 12.3 Incidence of radiation induced death and abnormalities in mice related to period at which irradiation (200R) takes place.
A comparative scale of human development is given. (From Russell, L. B. and Russell, W. L., 1954).

growth in this respect are pre-implantation, organogenesis and fetal development (Fig. 12.3).

During the period of pre-implantation (0–9 days in Man) the organism is particularly sensitive to radiation, the LD50/30 in mice being one-third that of the mature animal. Irradiation at this stage usually results in death and those organisms that survive may show little abnormality, except those having sex chromosome aberrations.

It is during the period of active organogenesis (9–42 days) that irradiation will produce severe anatomical malformations. During this phase of development in which major changes occur quickly in the organisation of the embryo the time of irradiation very much determines the nature and severity of the malformation. Although irradiation at this stage will normally result in severe structural abnormalities, high doses (in the range of LD50 to the mother) will prove lethal to most of the embryos (Table 12.1).

During the succeeding weeks of fetal development the incidence of anatomical defects after irradiation decreases except in the brain, eye and gonads which differentiate relatively late. At Hiroshima in children irradiated 'in utero' at this stage many had microcephaly and associated mental subnormality. Other malformations that occurred in these children were dislocation of the hip, heart disease and hydrocoele.

Table 12.1 Abnormalities that may follow appreciable irradiation exposure *in utero*.

Time (weeks) at which irradiated	Defect
0–4	Most resorbed or aborted.
4–11	Severe abnormalities most systems.
11–16	Mainly microcephaly, mental abnormality and growth retardation. Few skeletal, genital organ and eye abnormalities.
16–20	Few cases of microcephaly, mental sub-normality and growth retardation.
> 30	Unlikely to produce serious anatomical defects. May have functional disturbances.

Further information on the effects of intra-uterine exposure of the embryo and fetus has been obtained from studies of the Atomic Bomb survivors. At Hiroshima an assessment was made of 205 children who had been exposed to atomic radiations in the first half of intra-uterine life. 194 children were exposed at distances greater than about 1200 m and no abnormality was found; however, of the 11 children who were within this distance from the hypo-centre of the explosion seven had microcephaly and were mentally sub-normal. At Nagasaki it is reported that 30 pregnant mothers received irradiation within 2000 m of the hypo-centre, there were seven fetal deaths, seven neonatal deaths, and four who were mentally sub-normal out of the 16 surviving children.

It is extremely important to note also that irradiation of the embryo and fetus increases the risk of the child developing some forms of cancer or leukaemia and this is discussed in the sections on leukaemogenesis and carcinogenesis.

Avoidance of irradiation of embryo

Because of the risks related to irradiation of the embryo and fetus it is now advised that all clinicians make certain that a pregnant woman is not referred for abdominal X-ray examination, particularly during the first two weeks of pregnancy. At this time pregnancy may be unsuspected, but it is in this period that the most serious results of irradiation of the embryo will be seen. All clinicians are therefore asked to consider if a patient may be in the early stages of pregnancy and observe the '10-day rule' before requesting abdominal X-ray examinations. This rule recommends that, unless medical indications require it, X-ray examination of the lower abdomen should be carried out in the first 10-days following the first day of the menstrual cycle. In this way the possibility of irradiating the gravid uterus, even with a low dose is excluded.

Should the human embryo be irradiated even to a very low dose the question of therapeutic abortion should be considered. It has been suggested that abortion should be performed if the gravid uterus receives more than 10 rads, because of the high risk of congenital abnormality. Even with lower doses the possibility of abortion should be given careful consideration because of the risk of serious malformation.

Ageing and life-shortening

It has been observed in many experiments with animals that exposure to radiation accelerates the ageing process and usually will lead to a shortened life-expectancy compared with control animals.

The normal ageing process is poorly understood, but it is assumed to involve simply a progressive degeneration of normal tissues leading to functional impairment and eventually death. It is suggested that exposure to irradiation of

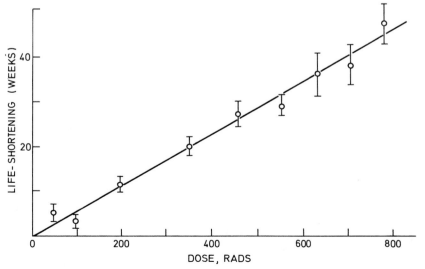

Fig. 12.4   Relationship in mice of shortening of life span to radiation dose. (From Rotblat, J. and Lindop, P., 1961).

the organism will result in depletion of stem-cells and differentiated cell populations and deterioration of the vasculo-connective tissues. In fact, the processes in the whole-body syndrome of premature ageing are similar to those involved in the pathogenesis of radiation changes in individual organs and tissues described in Chapter 10.

Rotblat and Lindop (1961) have demonstrated a linear relationship between life shortening and radiation dose (Fig. 12.4). A distinction may be drawn between the observations following chronic irradiation and single-dose exposures, after which the mortality rate does not change, but animals die at a younger age than controls. This effect is called precocious ageing. By contrast, after chronic irradiation the mortality rate of the exposed animals *is* increased, an effect described as accelerated ageing. In these animals it is seen that they develop degenerative illnesses at a much higher rate than their controls.

The reduction of life-span is thought then to be the result of the speeding up of degenerative processes, leading in addition in some cases to a higher incidence of specific diseases, such as cancer and early death. It has been shown by Rotblat and Lindop that the shortening of life span is greater when immature rather than mature animals are irradiated, although the effect increases again with age. (Fig. 12.5).

There is really no good evidence in Man that radiation has been responsible for shortening of the life span. A study of American Radiologists over the period 1945 to 1954 was conducted which appeared to demonstrate that, compared to other physicians, they had a reduced life span. A large number of assumptions had to be used in this retrospective review and there is considerable doubt about its validity.

In response to the report on the American radiologists a study of the life span of British radiologists in practice from 1897 to 1957 was conducted and this did not show any evidence of deleterious effect of occupational exposure to radiation. It is also true that there has been no evidence from survivors of the two atomic bomb explosions in Japan of any non-specific life-shortening effect, although there is an excess in the mortality from aplastic anaemia and various forms of cancer.

Lengthening of life span

It has been noted in some experiments when the dose rate was about 0.1 to 0.5 rads per day that the irradiated animals have lived longer than controls (Upton, 1960). The reasons for this observation are far from clear. It has been suggested that the effect may be due to environmental factors unconnected with the radiation. Certain effects produced by the low radiation doses have been suggested, such as increased activity of the lympho-reticular system and the slight increase in neutrophils and lymphocytes that has been observed, may be advantageous to the animals. Less convincing explanation of the beneficial effect have been that the radiation may inhibit the progress of coincidental diseases and also perhaps the development of new diseases. There is no evidence to suggest that this effect occurs in Man.

Aplastic anaemia

It has been suggested that excessive exposure to radiation will produce an increased incidence of aplastic anaemia in human subjects. The best evidence (Court-Brown & Doll, 1965) reported that in patients treated for ankylosing spondylitis with radiation above 100 rads there was an increased incidence of aplastic anaemia. It is important to record that among the survivors of the atomic bomb explosions no significant increase in aplastic anaemia has been observed. However, animal experimental evidence does support the opinion that radiation does produce aplastic anaemia, but the risk is much less than for the induction of leukaemia.

Radiation carcinogenesis and leukaemogenesis

Much concern is often expressed about the possible carcinogenic effects of ionising radiations and yet, compared to many chemical agents, these radiations are not highly dangerous in this respect. It is clear that the leukaemias are the most important neoplastic diseases induced by ionising radiations and their incidence in Man may be accounted for in part by exposure to radiation. It has been estimated that perhaps over half the other forms of cancer in Man are caused by chemicals, but the overall increase in incidence due to radiation is very small indeed.

The incidence of malignant change following irradiation is so small that large populations at risk have to be studied to allow reliable interpretation to be made of the observations. Since the latent period for the induction of cancer is so long, 5 to 8 years for leukaemia and perhaps 15 or more years for most other forms of cancer, these observations have also to be made for very many years in most circumstances. Uncertainty of the spontaneous incidence rates of cancer and leukaemia in particular populations and the incidence of deaths from intercurrent diseases also makes comparative analysis of data from exposed populations a difficult statistical exercise. The incidence of both cancer and leukaemia naturally increases with age and this factor must be carefully controlled in assessing the risks of radiation exposure. The International Commission for Radiological Protection has estimated that whole-body exposure to the adult population of 1 rad per year would result in 2 cases of leukaemia and 2 cases of other forms of cancer per 100 000 population per year. The total level of background radiation including natural and other sources in the United Kingdom is about 0.1 rad per year and the number of cancers produced by this order of dose is very small indeed (approximately 0.07 per cent) compared to the number of cases from other causes. In respect of leukaemia, however, it has been estimated that 10 per cent of all cases may be due to radiation exposure.

Exposure to ionising radiations certainly increases the natural rate of malignant transformations in some organs, but the amount by which the probability is increased in relation to the dose of radiation received is often uncertain and ill-understood. It is still not possible to exclude the existence of a threshold dose below which there is no increased incidence of cancer although this is thought to be unlikely. It has been thought that the incidence of cancer was a linear function of radiation dose, but this implies a single event transformation with dose. Such a hypothesis is not supported by most other evidence although some clinical and experimental data presented later are apparently consistent with this theory. It must be remembered that radiation also kills cells and that as the dose of radiation increases the probability of cell death becomes much greater than neoplastic transformation. Indeed it has also been suggested that the potential tumour cells may be more susceptible to being killed by high doses of radiation. Therefore the incidence of malignant disease will decrease again at high doses. The most reasonable dose-response relationship for the induction of cancer and leukaemia is illustrated by the curve of Fig. 12.5. In this

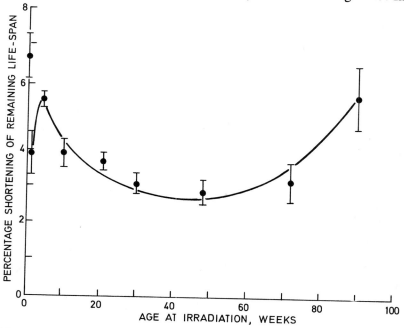

Fig. 12.5   Relationship in mice of shortening of life span to age at irradiation. (From Rotblat, J. and Lindop, P., 1961).

curve there is an initial ascending part rising to a summit before a final descending part. It is the accurate measurement of the initial part of the curve that has proved to be impossible to determine with animal experiments as, in this very low dose range (under 100 rads), unacceptably large numbers of animals would be required. Borek and Hall (1973) have developed an *in vitro* system of embryo hamster cells and have shown transformed clones of these cells to be produced after irradiation of only 1 rad X-rays. These cells may be readily recognised *in vitro* as being transformed and also will form fibrosarcomas when injected into young hamsters. Their dose-response curve from 1 to 600 rads follows a similar shape to the curve for murine myeloid leukaemia illustrated in Figure 12.5, and suggests a linear dependence of induction on dose (Fig. 12.6) at low doses. Uncertainty still exists about the precise dose-

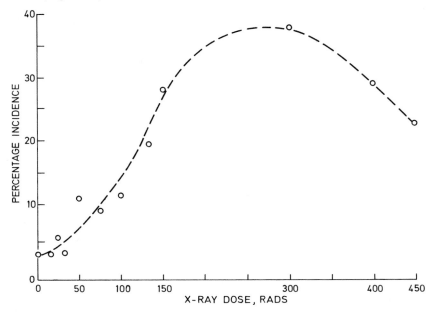

Fig. 12.6 Incidence of myeloid leukaemia in mice following whole-body irradiation. (From Upton, A. C. 1961).

effect relationship at low levels of exposure. Figure 12.7 illustrates three possible shapes of curves that may apply at very low doses. It will be seen that extrapolation from high-dose data will influence the assessment of risk at low doses, depending on which curve is considered to be most applicable.

It should be pointed out that the considerations of the International Commission on Radiation Protection are based on the hypothesis on a linear dose-response relationship and this assumption may over-estimate the hazard of doses below 100 rads.

It must be appreciated that it is not only the total dose of radiation that is important, but also dose-rate in relation to cancer induction. There is, however, little information available on the influence of dose-rate on the incidence of neoplastic transformation, but it does seem that higher dose intensities are more effective. The induction of leukaemia following radiation may be an exception as it may be independent of dose-rate. Fractionation of the dose may make the effects of radiation less pronounced, but the data available does not provide any confidence about the possible protection which prolongation of the total dose delivered may provide. Further experiments by Borek and Hall, have shown that split dose exposures to X-rays are *more* effective in producing *in vitro* transformation than single doses.

The quality of radiation has also to be considered and high LET radiations such as fast neutrons are much more effective in producing transformation than

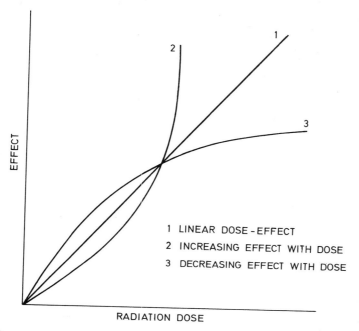

EFFECT

1 LINEAR DOSE-EFFECT
2 INCREASING EFFECT WITH DOSE
3 DECREASING EFFECT WITH DOSE

RADIATION DOSE

Fig. 12.7 Theoretical dose-effect curves at low doses for cancer and leukaemia induction.

low LET radiations. The clinical experience of Nagasaki and Hiroshima has suggested that an RBE of 5.0 was appropriate for leukaemia induction in respect of the neutron irradiation (compared to X-rays) released by the atomic explosion at Hiroshima. The RBE of high LET radiations is dose-dependent and some caution has to be exercised in assuming a single particular value for carcinogenesis and leukaemogenesis, especially also when the mechanisms involved may be different for different tissues. For example an RBE of 20.0 for fast neutrons has been reported for the induction of thyroid cancer in a certain strain of rat and more information is needed on these effects in other tissues and organs.

The nature of the essential injury that produces malignant transformation is unknown although much evidence points to damage of the DNA structure such as gene mutation. The demonstration that patients with Xeroderma pigmentosum who carry a high probability of developing skin cancer after exposure to ultra-violet radiation and who have defective repair of DNA-induced radiation lesions also supports this hypothesis.

Pathogenesis of radiation-induced neoplastic disease

The pathological processes involved may be considered to be based on either the local or abscopal effects of irradiation on the tissues concerned (Table 12.2) and possible mechanisms involved are now described.

Table 12.2 Pathogenesis of Radiation Induced Cancer and Leukaemia

1. Scar tissue instability	
2. Somatic mutation	LOCAL
3. Viral activation	
4. Immunological depression	
5. Cell population kinetic stimulation	ABSCOPAL
6. Endocrine disturbance	

Scar instability

It is almost 150 years ago since the first description of cancer of the skin arising on scar tissue following burns. Many other examples may be given of malignant transformation on scar tissue or areas of erythema *ab igne* and usually tissue damage following radiation is evident before cancer induction is seen. This may be a factor of small importance in the induction of cancer, but skin and other cancers, for example, may occur with greatly increased frequency in irradiated areas in which no gross radiation changes are observed.

Mutation hypothesis

An attractive hypothesis is that the transformation of an irradiated cell occurs as a result of an induced gene mutation. This is too simple an explanation although it may be that a single gene mutation may contribute to, or indeed initiate, the malignant process while other factors eventually promote the development of a cancer. There is some animal experimental evidence that demonstrates a linear relationship of carcinogenesis to dose to support the theory of somatic mutation, but the evidence cannot support this as a general hypothesis for the induction of cancer and leukaemia.

Viral activation

It has been demonstrated that the role of radiation in the induction of the thymic lymphomas, osteosarcomas and certain leukaemias in mice may be by activation of latent viruses (Kaplan, 1974). The mechanisms of interaction may be through damage to certain structures of the cell that leads to the release of the active virus. Certain other effects may also have to apply, such as concomitant injury to the cell population in which transformation takes place. These changes may produce either abnormal cell forms or, as in the case of murine leukaemia, maturation arrest in the process of differentiation that leads to the production of large numbers of immature cells of the lymphoid series. A number of other studies in mice have identified viruses, other than the Rad LV virus which induces lymphosarcoma and lymphatic leukaemia, that become activated to produce malignant transformation at the site of the radiation injury.

Immunological depression

It has also been suggested that radiation may be followed by cancer formation due to the depression of the immune system, either locally within a small field of irradiation or by general haematological depression resulting from wide field irradiation of the lymphoid and haemopoietic tissues. Changes of this kind may allow abnormal cells to proliferate and form a cancer that would otherwise have been identified and rejected. It is known that patients who are immunosuppressed, do have a signficantly increased likelihood of developing cancer, particularly of the lympho-reticular system. It has also been considered that immunological depression may allow proliferation of an oncovirus and indirectly be responsible for the establishment of the cancer process.

Cell population kinetic stimulation

The important research by Kaplan (1974) of the induction of leukaemia on the C57 Br and other strains of mice illustrated that activation in the kinetics of the target organ population was important. After irradiation, the thymus is depleted of cells but in response to this there is a rapid proliferation of thymocytes resulting in thymic hyperplasia. It was shown that this hyperplasia is essential in the induction process and that other changes such as maturation arrest in the thymus are also important indicating that alterations in the cell population kinetics of an organ may be antoher mechanism in the process of cancer induction.

In this respect certain cancers of the endocrine system produced both directly and indirectly by radiation may be regarded as particular examples of the effect of cell population kinetic stimulation.

Endocrine disturbances

The induction of cancer as a result of endocrine disturbances following radiation is a clear example of an abscopal, or indirect mechanisms by which these changes may be produced. It has been shown in the mouse that ablation of the thyroid gland by localized irradiation may produce tumours of the pituitary gland. It seems also that an overactive endocrine gland may be more susceptible to malignant change following radiation and the inter-relationship of radiation effect and pre-disposing functional derangement in the induction of malignant tumours of the endocrine organs is extremely complex.

The latent period

It is particularly interesting to note that there is always a long latent period of a number of years between the time of irradiation and the clinical appearance of the induced cancer. The processes that account for this delay are not known, but it may indicate that factors other than the irradiation are involved in the promotion of the neoplastic disease. The long latent period may simply be a reflection of the number of transformed cells and their cellular kinetics that leads to the slow accumulation of abnormal cells. The minimum latent period in Man for leukaemia is usually about 4-5 years and for cancer commonly over 10 years. The mean latent period for most cancers induced by radiation is about 15 years, while the mean latent period for the induction of leukaemia is only about 8 years. The period of risk continues much longer for cancer (> 25 years) than for leukaemia (about 20 years).

Leukaemogenesis and carcinogenesis in man

Much of the information on the carcinogenic effects of radiation in Man is incomplete and unsatisfactory as the surveys have always been undertaken retrospectively. Inferences have to be drawn about effects produced under inadequately documented circumstances and often with rather crude estimates being made of radiation dose received. Table 12.3 lists the sources of evidence available concerning radiation carcinogenesis in Man.

Table 12.3 Data on relationship in man of radiation exposure and leukaemogenesis and carcinogenesis

1.	*Environmental radiation exposure*	
i.	Atomic Bomb	Hiroshima
		Nagasaki
		Radioactive mine workers
ii.	Occupational	Luminous Dial Painters
		British and American Radiologists
2.	*Medical radiation exposure*	
i.	Diagnostic	Fetal
		Thorotrast
		Ankylosing Spondylitis
ii.	Therapeutic	Childhood irradiation 'thymus'
		Metropathia Haemorrhagia

Leukaemogenesis

Atomic bomb explosions

The survivors of the atomic bomb explosions in Hiroshima and Nagasaki provide the best evidence in Man of an increased incidence of both leukaemia and various forms of cancer following irradiation. It has been shown that in the estimated dose range of 100 to 900 rads there is a linear increase in the number of cases of leukaemia, mainly acute and chronic myeloid leukaemia. The incidence of leukaemia is higher and the slope of the incidence line greater in survivors from Hiroshima who were exposed equally to γ-rays and neutrons, compared to Nagasaki, where the explosion produced predominantly (90 per cent) gamma radiation. At low doses there is some doubt about the relationship of leukaemia induction to radiation dose. In Hiroshima where there was mixed gamma and neutron radiation there appears to be no threshold value and small doses considerably increased the incidence of leukaemia. From the Hiroshima data the mixed radiation would seem to have increased the incidence of leukaemia by 3 cases per million population for every 1 rad of exposure. At Nagasaki there appeared to be a threshold dose of about 50 rads. The minimum latent period for induction of leukaemia was just under 3 years with a peak incidence between 4 and 8 years. After this period of observation the incidence of new cases declines, until after 13 years the incidence of leukaemia is a little more than the expected number (Fig. 12.8).

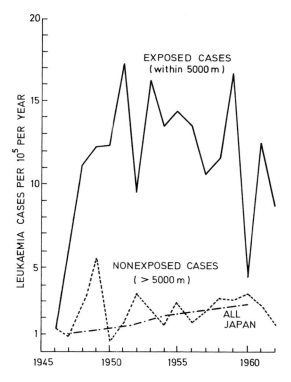

Fig. 12.8   Incidence of leukaemia in Hiroshima. From U.N. Scientific Committee on the effects of atomic radiation (1964). Ninteenth Session: Supp. 14 (A/5814).

Occupational exposure

It has been reported that American radiologists in practice between 1948 and 1961 had an increased mortality from leukaemia, three times that of the control population. There was also an excess of deaths from myeloma and aplastic anaemia, but there was no record of the size or nature of radiation exposure received by these radiologists over their lifetime. These findings were not substantiated by a later study of British radiologists.

Ankylosing spondylitis patients

In 1965 Court-Brown & Doll published the results of a survey of 13 352 patients (of whom 11 287 were men) who had been treated in 81 British radiotherapy centres between 1935 and 1954. Most of these patients had been given orthovoltage irradiation to the whole spine in two to four courses over a variable period of time. The mean radiation dose to the bone marrow was estimated and correlated with the observed incidence of leukaemia and compared with the expected rate in the population. It was not possible to relate the incidence of leukaemia to dose below about 300 rads, but from that level to about 1500 rads a linear relationship seemed to be consistent with the data. In this dose range it was estimated that the increased risk of leukaemia was 1 case per million male population per rad exposure. In this series of patients with ankylosing spondylitis 84.5 per cent were young males and it is also known that the haemopoietic system is not normal in many of these patients, and so one may not make general conclusions about the risks of radiation leukaemogenesis in the normal population. Figure 12.9 shows the increase in

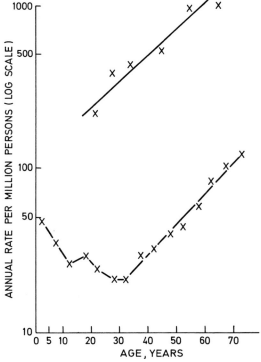

Fig. 12.9 Death rate from leukaemia (other than chronic lymphatic leukaemia) by age in British males (lower curve) and in British male patients given radiotherapy for Ankylosing Spondylitis. (Upper curve). Note natural increase in incidence with age. (After Doll, 1963).

both the natural incidence and radiation-induced leukaemia in British males. Irradiation appears simply to increase the spontaneous incidence of this group of diseases. It should be recorded that the minimum latent period to diagnosis following irradiation was 2 years with a peak incidence at 3 to 5 years. The increased incidence had almost disappeared by the 12th year after exposure.

Children irradiated in utero

The first report suggesting that small doses of radiation to the embryo and fetus increased the incidence of leukaemia and other cancers was published over twenty years ago. Since then there have been many conflicting reports principally because of the difficulty in obtaining satisfactory control groups. Stewart and Kneale (1970) have, however, produced the results of a more detailed study of *in utero* irradiation in relation to the number of radiographs taken. Together with MacMahon (1962) in the United States they have confirmed the increased risk of neoplasia in irradiated children *in utero*. An important finding of Bross & Natarazan (1972) is the ability to identify high risk groups amongst children who have been subjected to irradiation *in utero,* that depends on the incidence of other diseases, such as certain viral infections in the mother or child.

Treatment of thyroid disease

An increase in the incidence of leukaemia has been recorded in patients who have been treated with high doses of radio-iodine for cancer of the thyroid. It has been ascertained that the much lower doses of radio-iodine given in the treatment of thyrotoxicosis has not, in the detailed follow-up of a large number of patients, been shown to increase the incidence of leukaemia.

There is no doubt that high doses of ionising radiation of all types will increase the incidence of leukaemia, but there remains doubt about the hazard related to low-dose exposure.

Carcinogenesis

Atomic bomb explosions

The considerable increase in the incidence of many forms of cancer in survivors from the Japanese atomic bombs is well documented. As was seen in the case of leukaemia, the incidence of cancers increases with the estimated radiation dose and at the highest dose levels, within 1000 m of the hypocentre the number of observed cancers was four times that expected in the population. An increase in cancers of the breast, thyroid, lung and salivary glands have been reported and relatively small doses of 50 rads considerably increased the incidence of breast and thyroid cancer in survivors of both Hiroshima and Nagasaki.

Occupational exposure

The two classical examples to which reference is always made are the lung cancers induced in miners of radioactive substances and the osteosarcomas which developed in the jaws of the watch dial painters who used the luminous mixture of radium, mesothorium and radiothorium.

The high incidence of lung cancers in radioactive mine workers in Colorado, U.S.A. and in Jachymov and Schneeberg is well established. These workers who were extracting uranium and pitchblend were exposed for many years to very high doses by the inhalation of the radioactive gases, estimated to be about 3000 rads in many cases. The average latent period was about 15 years. It should be noted that an increase in lung cancer has also been observed in survivors of the Japanese Atomic Bomb explosions.

Thirteen workers from a factory where the faces of watches were painted with a radioactive paste were found to have developed osteosarcoma of the jaw after about 25 years. In addition shortly after the introduction of radioactive substances many patients were given radium salts by mouth for a variety of ailments and a proportion developed malignant disease of the skeleton.

Use of thorotrast

Thorotrast, a suspension of thorium dioxide, was an excellent contrast medium used in diagnostic radiology until about 1945. Thorotrast, unlike radium, is deposited mainly in the lympho-reticular system, where it is retained. The effect on the liver producing intense fibrosis and then haemangio-sarcoma about 18 years later has been described in Chapter 11. In a few patients cancers developed at the site of the injection of the radio-active material, usually associated with a severe fibrotic reaction.

Ankylosing spondylitis patients

Ankylosing spondylitis patients (Court-Brown and Doll, 1965) have a samll but significant increase in the numbers of cancers developing in the irradiated mid-line structures. The increased risk of cancer, after the first five years, is about 60 per 100 000 per year and appears to increase with time. This finding is unexpected as the incidence should 'plateau' and may be explained by some patients receiving further irradiation for their spondylitis.

Thyroid irradiation in childhood

A very high incidence of thyroid cancer has been found in children who had irradiation of the neck or mediastinum in infancy for enlarged thymus or other benign conditions. The period of high risk of induction of thyroid cancer may continue to the age of 10 to 15 years and thereafter there is a long latent period of probably about 15 years.

Children irradiated in utero

It has been estimated by Stewart and Kneale (1970) that 1 rad delivered to the fetus might result in an increase of 572 ± 133 cancer deaths per million before the age of 10 years. MacMahon (1962) in the United States has estimated in a survey of children born between 1947 and 1954 that intra-uterine irradiation was responsible for a 40 per cent increase in mortality from cancer.

Treatment of metropathia haemorrhagia

Smith and Doll (1976) have assessed the risk of leukaemia and of developing cancer in the pelvic organs following ovarian irradiation for metropathia. They have observed that from the fourth to the nineteenth years following exposure the risk of leukaemia is increased three-fold. There is also a small increase (1.5) in cancer in the heavily irradiated area, very similar to the effects seen in irradiated patients with ankylosing spondylitis.

Effects of high-dose radiation in cancer therapy

It should be noted that the data on radiation leukaemogenesis and carcinogenesis in Man are derived from low or moderate dose exposures. There is no evidence to suggest that the very high dose levels given in cancer therapy increases the incidence of leukaemia or further cancer in successfully treated patients. It is important to be aware that a large series of 29 493 women treated for cancer of the uterine cervix by intra-cavitary radium or external beam therapy have shown *no* increase in leukaemia or cancer incidence. (Hutchison, 1968). High doses of radiation are likely to cause cell ceath rather than malignant transformation and it may be that at the dose levels of radiation used in cancer therapy the risks of leukaemia or cancer induction are not greatly increased.

The benefits of judicious radiotherapy, for certain benign conditions as well as in the management of cancer, certainly far outweigh the very small risk of radiation-induced neoplasia.

Principles of radiation protection

It will now be understood that at low doses of radiation exposure the hazards may be either genetic or somatic.

The *genetic risks* are extremely difficult to evaluate and little data are available on the genetic effects of radiation in Man. The International Commission for Radiation Protection (I.C.R.P., 1966) has recommended that 20 rads should be considered the mutation doubling dose, that is, the dose of radiation that would double the natural incidence. It is assumed for this standard, that the population exposed to 1 rad would expect an increase in mutations of $1/_{20}$ of the expected spontaneous mutation rate. The genetic detriment will be manifest in a number of ways in the form of autosomal and sex-linked gene characters, chromosomal aberrations, abortions and what are called genetic deaths. Genetic death implies the loss of a particular cell line either as a result of premature death or impaired fertility. The genetic effects will be most pronounced in the first generation after irradiation, but the total number of deleterious effects of genetic mutation following radiation will be much greater in all the subsequent generations. It must be made clear that those effects that may be the result of small doses of radiation are extremely rare compared to the natural mutation rate; and indeed, many environmental pollutants and many cytotoxic drugs are possibly now responsible for much greater genetic damage than radiation.

The somatic risks have been described in some detail, that is the developmental abnormalities, life-shorteneing and carcinogenesis.

It has been shown that quite small doses of radiation may induce malformations in the embryo and fetus and the I.C.R.P. (1966) have recommended the observation of what is known as the *10-day rule*. It should be remembered that the risk of developing leukaemia is increased about five-fold when the radiation dose is received *in utero*. As far as the effect of small doses of radiation on life expectancy is concerned I.C.R.P. feels that the available data do not justify making a quantitative estimate of risk.

The data on the incidence of cancer in humans are sound, for high doses, but the evidence is not available to make estimates of risk with doses of less than 100 rads. Since the high dose data appear to be linear it has usually been the practice to assume that a linear relationship of risk to dose may hold for low doses. This does mean that the estimates of risk may be over-estimating the real hazard potential. Such calculations that relate to the annual risk to human objects when subjected to 1 rad continuous radiation exposure each year, or to a lifetime risk (or rather in the 20 years) after a single dose of 1 rad, estimate 20 cases of leukaemia and 20 other cancers per million population exposed.

On the basis of these considerations which have been briefly mentioned the International Commission on Radiological Protection (I.C.R.P.) make recommendations of the *maximum permissible dose* levels for the population that may be exposed to radiation. The levels are determined to minimise the incidence of genetic effects and deleterious somatic effects and those maximum permissible doses should be kept as low as possible.

It is recommended that the population should normally not receive a whole-body dose greater than 0.5 rem per year over a period of 30 years, excluding background and medical exposure. For the population who are occupationally exposed to radiation, the standards have to be less stringent. The dose to the

Table 12.4   Maximum permissible doses (rem per year)

Tissue	Occupationally Exposed	General Population
Whole-body, gonads, bone marrow	5	0.5
Skin, thyroid* bones	30	3.0
Hands, wrists, feet, ankles.	75	7.5
Other organs	15	1.5
Total exposure (30 years)	5 (Age–18)	5.0

*Children under 16 years dose to thyroid is limited to 1.5 rem/year.
Tissue dose in rems = absorbed dose in rads × RBE.

whole-body, gonads and bone marrow, must not exceed 5.0 rem per year and certain organs may receive considerably higher doses. In each case the dose is ten times higher than is permitted for the general population, as shown in Table 12.4, but the gonad dose is still four times lower than the estimated genetic doubling dose of 20 rads.

Detailed recommendations for practice in the United Kingdom may be found in the *Code of Practice for the use of Ionising Radiations* (1972). Observation of the advice given in this publication (that is based on the I.C.R.P. reports) will ensure that the hazards of radiation are well controlled. Nevertheless, all radiation exposure is potentially harmful and it is essential that the unnecessary irradiation of individuals and the general population is kept at minimal levels, so that the undoubted benefits of ionising radiations may continue to be realised with greatest possible margins of safety.

REFERENCES

Amorose, A. P., Plotz, E. T. & Stein, A. A. (1967) Residual chromosomal aberrations in female cancer patients after irradiation therapy. *Experiments in Molelcular Pathology,* **7,** 58-62.

Borek, C. & Hall, E. J. (1973) Transformation of mammalian cells 'in vitro' by low doses of X-rays. London. *Nature,* **243,** 450-453.

Bross, I. D. J. & Natarazan, N. (1972) Leukaemia and low-level irradiation. Identification of susceptible children. *New England Journal of Medicine,* **287,** 107-110.

Code of Practice for the Protection of Persons against Ionising Radiations arising from Medical and Dental Use. (1972) London: H.M.S.O.

Court-Brown, W. M. & Doll, R. (1965) Mortality from cancer and other causes after radiotherapy for ankylosing spondylitis. *British Medical Journal,* **2,** 1327-1332.

Doll, R. (1963) Interpretations of epidemiological data. *Cancer Research,* **23,** 1613-1623.

Hutchinson, G. B. (1968) Leukaemia in patients with cancer of the cervix treated with radiation. *Journal of the National Cancer Institute,* **40,** 951-982.

International Commission on Radiological Protection. (1966) The evaluation of risk from radiation. C. R. P. Publication 8. Oxford: Pergamon Press.

Kaplan, H. S. (1974) The role of radiation in experimental leukaemogenesis. *National Cancer Institute Monographs,* **14,** 207-220.

MacMahan, B. (1962) Pre-Natal exposure and childhood cancer. *Journal of the National Cancer Institute,* **28,** 1173-1191.

Revell, S. H. (1955) Chromatid aberrations in root-meristem of *Vicia faba. Mutation Research,* **3,** 34-53.

Rotblat, J & Lindop, P. (1961) Long-term effects of a single whole-body exposure of mice to ionising radiations II. Cause of death. *Proceedings of the Royal Society — B* (London), **154,** 350-368.

Russell, L. B. & Russell, W. L. (1952) An analysis of the changing radiation response of the developing mouse embryo. *Journal of Cellular and Comparative Physiology,* 1: **43,** 103-149.

Smith, P. G. & Doll, R. (1976). Late effects of X-irradiation in patients treated for metropathia haemorrhagica. *British Journal Radiology,* **49,** 224-232.

Stewart, A. & Kneale, G. W. (1970) Radiation dose effects in relation to obstetrics, X-ray and childhood cancer. *Lancet,* **i,** 1185-1187.

United Nations Scientific Commission on the Effects of Atomic Radiation. (1964) Nineteenth Session, Supp. 14 [A/5814].

Upton, A. C. (1960) Ionising radiation and aging. *Gerontologia,* **4,** 162-169.

Upton, A. C. (1961) The dose-response relation in radiation induced cancer. *Cancer Research,* **21,** 717-729.

FURTHER READING

UNSCEAR (1972) Ionising Radiations: Levels and Effects. Vol. **2,** Effects. United Nations. New York. A Report of the United Nations Scientific Committee on the Effects of Atomic Radiation to the General Assembly.

13. Fractionation

The practice of giving multiple small daily fractions of radiation was first introduced in an attempt to irradiate as many tumour cells as possible during mitosis since this was recognized as the most sensitive phase of the cell cycle. The use of multiple fractions had also been shown to spare overlying normal tissues while causing severe damage to relatively deep seated structures. The early demonstration of this fact by Regaud following experiments on the rat testes, was confirmed by pioneer radiotherapists who found that fractionation of the total dose of radiation had a relatively favourable effect on normal tissues while still having a lethal effect on tumours. The object of fractionation in radiotherapy is to kill all tumour cells without producing serious damage to the surrounding normal tissues which necessarily must be included in the volume of high dose irradiation.

It is now clear that the clinical benefits of giving multiple dose fractions results from the complex interactions of many biological and physical factors (Tables 13.1 and 13.2). The cumulative effects of these factors will depend on

Table 13.1 Biological factors of importance in fractionated radiotherapy

1. Intrinsic radiosensitivity of cells
   (a) Cell population may not have uniform sensitivity.
2. Repair of sublethal radiation damage.
3. Oxygenation of cells,
   (a) Proportion of anoxic cells
   (b) Reoxygenation.
4. Repopulation
   (a) Differences between tumour and normal tissue
   (b) Changes in proliferation rates following irradiation.
5. Mitotic delay.
6. Redistribution of cells in cycle.
7. Potential radiation damage.
8. Non-lethal damage.

Table 13.2 Physical factors of importance in fractionated radiotherapy

1. Quality factor
   linear energy transfer
2. Volume factor
3. Dose-time factors
   fractionation
   frequency
   prolongation
   periodicity
   protraction

the site, size and histological type of tumour, in addition to the radiotherapy technique employed, but there is as yet little evidence about their quantitative importance.

The number of variations in treatment schedules which may be found in clinical practice illustrate the gap which still exists between the science of radiobiology and the art of clinical radiation therapy. Indeed the practice of radiotherapy has been built up empirically by detailed and carefully collated clinical observations. Radiobiology still cannot provide a complete explanation for all phenomena which are observed clinically and is unable for reasons which will be discussed to provide, as yet, a fully rational scientific basis for fractionated radiotherapy.

Biological factors

The clinical effects of fractionated radiation are influenced primarily by the four 'Rs' of radiobiology described in some detail in Chapter 7. These are:
1. The capacity of mammalian cells to *recover* from sublethal damage.
2. *Repopulation* of the tumour and normal tissues between fractions.
3. *Re-oxygenation* of the tumour during course of treatment.
4. *Re-distribution* of normal and tumour cells in the cell cycle.

Another factor which may influence the clinical response to radiotherapy is the intrinsic radiosensitivity of the cells, that is their Do values, although this is not thought normally to contribute to any great extent to the differential effect which is obtained between tumour and normal tissues.

These factors and the others listed in Table 13.1 influence the probability of cell killing on which the therapeutic radiation response essentially depends. The clincial effects observed in normal tissues are also determined by the pathological processes of restitution which may be seen after the first dose of radiation. The gross clinical effects seen during the regression of tumours following irradiation depends not only on the degree of cell depletion but also on processes such as phagocytosis and perhaps, in some cases, immune mechanisms.

Intrinsic radiosensitivity

Although a range of values for the mean lethal dose (Do) has been demonstrated for a large number of different mammalian cells there is no consistent difference between normal tissue and tumour cells (Table 13.3), and (see Ch. 6). Indeed there is commonly found to be a close similarity in the mean lethal dose in cells of the same generic type whether they be normal or neoplastic. It can be seen that the range of values extend across quite different tissues. It cannot be claimed therefore that radiotherapy succeeds as it often does in the cure of cancer because the tumour cells are intrinsically more sensitive to the processes of radiation injury than normal cells. For the same reason one may not, on the evidence of Do values, select tumours which could be successfully treated by X-rays.

It must be recognized that most of these values are obtained by *in vitro* experiments or *in vivo* cell dilution techniques and the cells will not respond exactly in the same way when irradiated repeatedly as cells in an intact tissue in

Table 13.3 Cell survival curve parameters for some mammalian cells

Cell type	Do (oxygenated)	Dq
Mouse		
haemopoetic stem cell	100	100
leukaemia	100	115
mammary carcinoma	340	230
fibroblasts	200	210
skin epithelium	135	350
osteosarcoma	—	280
melanoma	85	280
intestine	130	450
stomach	—	550
Rat		
capillary endothelium	170	340

the live animal. This does not alter the principle already stated that there is no consistent difference in the intrinisic radiosensitivity which may explain the curative results of radiotherapy. It should be recognized, however, that small differences in cell survival for a given X-ray dose will be magnified during the course of fractionated radiotherapy and the contribution of small differences found in single dose experiments may have a greater effect than often believed (Fig. 13.1). The differential effect of repeated small doses of X-rays is more related to the shoulder region, than to the exponential part of the survival curve and in these circumstances the Do is not relevant. Indeed much greater differ-

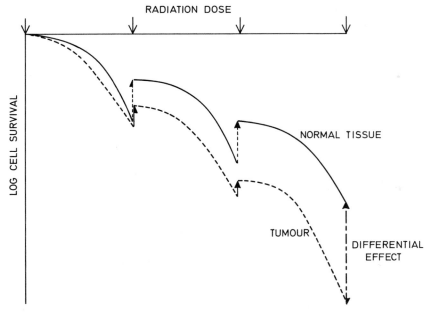

Fig. 13.1 Schematic representation of cell survival curves illustrating the increasing differential effect on tumour and normal tissue with fractionation. Small differences in survival curves will be amplified by fractionation.

ences between cell types are seen in the shoulder region than in the exponential region of mammalian cell survival curves.

It cannot be completely denied that in certain clinical situations, differences in radiosensitivity may contribute to the apparent resistance of the responsiveness of tumours compared to normal tissues. The major benefits of fractionated radiotherapy are to be explained by looking at the more subtle inter-related biological factors.

Recovery from sublethal radiation damage

Data from *in vitro* experimental results do not demonstrate any consistent difference in repair capacity of tumour cells and normal tissues, measured by the quasi-threshold dose Dq (Ch. 7). A wide range of Dq values ranging from 100 rads to 550 rads have been recorded in a large series of mouse experiments, some of the results of which are given in Table 13.3.

Many tumours are known to be hypoxic and diminished capacity to repair sub-lethal radiation damage has been reported in mammalian cells which are extremely deficient of oxygen, less than 10 parts per million. Cells which are less severely hypoxic, (<300 ppm) have been shown to have an unimpaired capacity for short-term recovery of sub-lethal radiation damage. If this is generally true then the accumulated damage of fractionated irradiation might be somewhat greater in tumours than in the surrounding normal tissues that are well oxygenated.

Little is known about the values of the shoulder or Dq in human tumours or normal tissues, but it does seem that it is a parameter of great importance in the total effect of a fractionated course of radiotherapy. For example, the survival curves obtained by irradiating cells of a mouse melanoma have been shown *in vitro* to have a very large shoulder and it was thought that this factor might explain the unfavourable results obtained clinically in the radiotherapy of malignant melanoma. However, studies on human malignant melanoma have not demonstrated a similarly large shoulder on survival curves from this type of tumour cell. It is likely to be the detailed shape of the shoulder region, as well as the maximum repair capacity, that determines the effect of multiple small fractions.

It is also suggested that some other tumour cells such as found in malignant lymphoma may have little or no capacity to repair sublethal damage and this may in part explain how effectively they may be managed by radiotherapy.

Great differences are found in the ability of normal cells to repair sub-lethal damage. Haemopoetic stem cells show little repair capacity while the epithelial cells of the skin and gastro-intestinal tract and the cells of growing cartilage have a large capacity to repair sublethal damage.

Repopulation

In the ideal clinical situation one would expect that normal tissues would be completely repopulated while the tumour showed no growth between fractions of the X-ray treatment. If that were so, the tumour would be progressively depopulated while the surrounding normal tissue would be maintained in a steady state. In practice there is possibly a beneficial effect of this kind to be found operating in many tissues which are included in the field of irradiation during fractionated treatment. It is true that there is no general difference in cell cycle times between tumours and normal tissues. The effective doubling times of tumours and of normal tissues may show considerable differences due to the effects of cell loss and other factors influencing the cell population kinetics (see Ch. 5).

In some normal tissues following the initial radiation injury there may be a surprisingly large increase in the proliferation rate until the normal tissue is fully reconstituted. This effect may be seen in the response of the haemopoetic stem cells where the response is immediate and the growth fraction increases together with a reduction in maturation rate so that the pool of precursor cells is renewed quickly and there may be an over-production of leucocytes, for example, for a time after X-ray treatment to a large volume of bone marrow. On the other hand the tumour will continue to show a high cell loss after each division although its growth fraction may enlarge for nutritional reasons, in order to compensate for the progressive cell depletion following successive doses of radiation as may occur in normal tissues. If such a differential depopulation were to be produced in tumours compared to normal tissues this effect would contribute greatly in improving the therapeutic ratio. (Fig. 13.2).

In tumours and in some normal tissues the cell cycle time may be shortened following irradiation so that the recurrent growth of the tumour may appear to be more rapid than would have been expected on the basis of the growth parameters determined before irradiation. Similarly in normal tissues such as

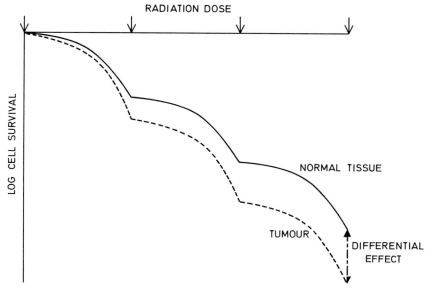

Fig. 13.2 Schematic representation of the cell survival curves illustrating the effect of differential repopulation of tumour and normal tissues during fractionation.

haemopoietic stem cells and the epithelial cells of the gastro-intestinal tract or skin irradiation may be followed by more rapid restitution of the depleted population of cells.

Animal experimental work has shown that in tissues such as the lung and in bone callus no cellular proliferation could be found up to one month after a large dose of irradiation. This, of course, may be attributable to the long turnover times of the vascular endothelium and connective tissue elements in these structures. It has been pointed out in Chapter 7 that even in skin, a tissue in which repopulation does occur during fractionated radiotherapy it is much less important than the factor of 'recovery' from sublethal radiation damage.

Reoxygenation

An important process which occurs in most tumours after irradiation is an improvement in oxygenation of the tumour cells. A possible sequence of events is illustrated in Figure 13.3 and was proposed by Thomlinson (1967). It is

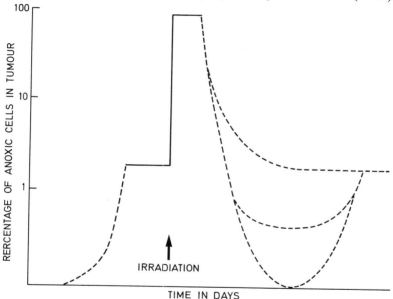

Fig. 13.3 Schematic representation of the proportion of viable anoxic cells in a tumour during its growth and following irradiation.
(After Thomlinson, R. H., 1967).

assumed that as a tumour grows an increasing proportion of its clonogenic cells will become anoxic until a plateau is reached at 10 to 20 per cent of the population. Immediately following irradiation there will be an increase in the proportion of hypoxic cells in the remaining viable tumour since the radiation will have killed a much greater proportion of well-oxygenated cells. Oedema too may adversely affect the micro-circulation through the tumour and for a time an increased number of cells will have an impoverished blood supply and be relatively more poorly oxygenated. The proportion of anoxic cells may reach almost 100 per cent. Thereafter the availability of oxygen will improve and the proportion of anoxic cells in the tumour diminish as a result of a number of inter-related effects. Death of the lethally irradiated population will quickly result in a decrease in the oxygen consumption of the tissue and as the killed cells are removed, the relationship of surviving cells in the capillary vascular bed will be improved, increasing the availability of oxygen to the remaining tumour cells. Also as the tumour shrinks in size with removal of dead cells and resorption of oedema, the blood flow in the tumour may improve with the decrease in tissue tension. Stagnant anoxia is known to be an important cause of poor oxygenation in many tumours. It has been demonstrated, too, that after irradiation there may be an increase in the vascularization of a tumour that will improve cellular oxygenation by decreasing the inter-capillary distance if the circulation is maintained. The proportion of anoxic cells may return to the pre-irradiation level or even to lower levels if oxygenation is much improved.

The improved oxygenation of the tumour will continue until the cells begin to replicate again and re-growth occurs. The number of anoxic cells in the tumour will then steadily increase until the proportion relative to well-oxygenated cells is reached, which is a characteristic of particular types of tumours. The anoxic fraction will depend on the morphology of the tumour and the functional activity of its cell population, but a commonly reported proportion of anoxic cells is 15-20 per cent (Hewitt *et al.*, 1967).

The temporal pattern of re-oxygenation has been demonstrated for a few experimental animal tumours (Fig. 13.4), and was discussed in Chapter 10, but nothing is known about the time course of events in human tumours. It can be seen that the pattern and degree of re-oxygenation varies in different types of

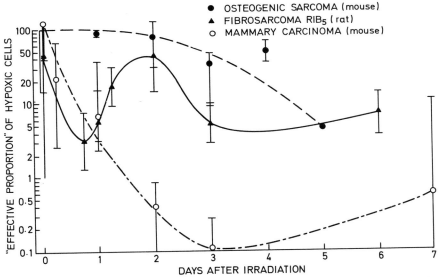

Fig. 13.4 The temporal pattern of re-oxygenation of three experimental animal tumours after a high single dose of radiation.
(From Thomlinson, R. H., 1969).

tumours, and this may be further influenced by the size and spacing of fractionated doses of irradiation during treatment.

Redistribution of cells in cycle

It has been explained in Chapter 6 that the radiosenstivity of mammalian cells varies throughout the cell cycle. The influence that redistribution of cells into other phases of the cell cycle will have in determining the response of tumours and normal tissues depends on a number of other biological and physical factors. The importance of this factor will depend primarily on the proportion of cells in the tumour or tissue that are actually in the mitotic cycle. It will also depend on the relative distribution of cell cycle times. The degree of variation of radiosensitivity throughout the cell cycle does depend on the type of mammalian cell and the influence of this factor can only be of importance when the dose-response variation is high.

It must be considered that when there is a significant differential radiosensitivity in the phases of the cell cycle, fractionated radiotherapy will produce a highly selective depopulation of the most sensitive cells leaving an increasing proportion of relatively resistant cells. The kinetics of the cell populations following irradiation will then critically determine if the residual population of cells will progress through the cycle to a more or less resistant phase when the next X-ray treatment is given (Fig. 13.5). The clincial effect will therefore

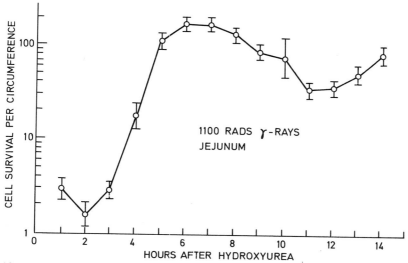

Fig. 13.5 The great difference in the survival of mouse intestinal crypt cells that follows exposure of 1100 rads of γ-rays at various time intervals after treatment with hydroxyurea, causing an accumulation of cells at the beginning of S phase. (From Withers, R. H., 1975).

depend not only on the level of dose delivered at each fraction, but equally on the interval between treatment fractions. (Withers, 1975).

In the normal circumstances of fractionated radiotherapy it is unlikely that even partial synchronization of the relevant cell population will be produced. It follows that it is unlikely that any signficant differential effect of this kind will be obtained that will increase the depopulation of the tumour compared to normal tissues.

Mitotic delay

All cells will be delayed by irradiation to some extent in proceeding to division and this block in the cell cycle is influenced by the size of the radiation dose and also by the age of the cell at the time of irradiation (see Fig. 4.3 Ch. 4). Mammalian cells will accumulate in the phase of the cycle just before mitosis known as G_2, which is a period of increasing radiosensitivity. The delay amounts to approximately one cell cycle time for every 1000 to 1500 rads, from 10-hour cycles *in vitro* to 100-hour cycles in skin. When one considers the size of dose fractions normally given in radiotherapy it can be appreciated that this effect is probably of little importance in determining the clinical end-result.

Potentially lethal damage

It has been established that mammalian cells in stationary phase (Go) of the cell cycle when cultured at 37°C may recover from what would otherwise have been lethal damage following irradiation. This effect, which is called potentially lethal damage, is dose dependent and is distinct from sublethal radiation damage which only occurs in cycling cells (see Fig. 7.4, Ch. 7). This does mean that cells that are not in cycle at the time of irradiation are capable of shedding this particular type of injury and are in effect a more radio-resistant population of cells than cells in other phases of the cell cycle. They could be the origin of recurrent cancer after fractionated radiotherapy if their number accounted for a high proportion of the tumour during the course of treatment. This effect is dose-dependent and although of real importance after high dose fractions, it is likely to contribute little to the biological damage which follows the relatively low doses of radiation usually given in a conventional fractionation regime.

Non-lethal damage

After large doses of radiation it is known that the numbers of cells in some surviving clones increase much more slowly than in unirradiated proliferating cells. This effect has been called non-lethal radiation damage and it has been suggested that it may result in tumours regrowing more slowly after irradiation than would otherwise have been expected.

While it is true that these slowly growing surviving cells are more radiosensitive than proliferating normal cells this fact is of little importance in clinical practice. As the population of cells in the surviving clones increases, the relative proportion of slow-growing cells with non-lethal radiation damage rapidly decreases and so their importance in determining the final radiation effect on a tumour or normal tissue is of very minor significance in fractionated radiotherapy.

Physical factors

The clinical response after radiotherapy is also determined by a number of physical factors (Table 13.2) which influences the relative importance of the biological effects which have just been described.

Radiation quality

The biological effect is determined by the quality of the beam of radiation. Quality of orthovoltage X-ray beams is usually defined by specifying the peak generating voltage and the half-value thickness of the effective treatment radiation; with megavoltage radiation only the peak generating energy is given as half-value thickness is too crude an index with such penetrating radiations. These terms are found to be inadequate to describe beams of radiation used in scientific work in which the biological effects are related to the primary ionising events and their spatial distribution. Attempts have therefore been made to define radiation quality in terms of the biological response and to consider radiation to be of the same quality when their biological effect is the same for each unit of absorbed dose.

The absorption of radiation beams in tissues may be regarded as taking place in discreet events and the biological effect is thought to depend on one, or at most a few, of these events. The rate at which these secondary charged particles deposit energy in the tissues in unit distance is known as the *Linear Energy Transfer* expressed as keV/μm. Although it is accepted the LET is a reasonable measure of radiation quality it is still not possible to estimate its value in absolute terms. This is because LET is specified as the *average* rate of energy loss along the tracks of the ionising particles in the tissues. A major difficulty is that most radiation beams are made up of a spectrum of radiation energies and measurement of the average energy loss provides a rather crude estimate of the effective quality of the radiation (see Ch. 2).

The *relative biological effectiveness* of radiation can be correlated with the estimates of LET values as shown in Figure 13.6. It should be noted that as the LET increases above 10 keV/μm it is not possible to give a single value for the RBE. Beyond this LET the shape of the cell survival curves becomes significantly different in the shoulder region than with low LET radiation, and the RBE depends on the level of biological effect produced. The greater the effect or the larger the radiation dose the smaller the RBE and vice versa. The RBE increases to a maximum value at approximately 100 keV/μm and then rapidly decreases. This is explained in that, as the density of ionization increases, so the probability increases of producing the required number of events to be lethal to the cell. When the LET is greater than about 100 keV/μm energy is 'wasted' by producing excessive ionization in the critical target. Although it is of fundamental importance to note that the RBE decreases because of 'overkill' with radiations of LET over 100 keV/μm, it is unlikely that radiations of this quality will be generally suitable for clinical application. The quality of radiations of clinical interest lie in the LET range of 0.3 keV/μm

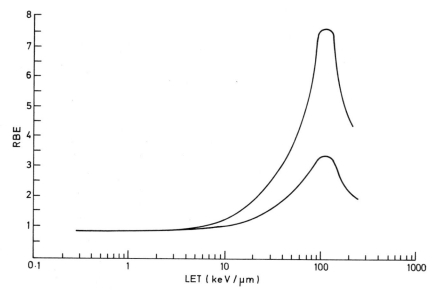

Fig. 13.6 The relationship of RBE to LET.

(Telecobalt) up to about 20 keV/μm average (Fast neutrons). Each of these ranges has a wide spread however. The LET of fast neutrons for example covers the range from 0.2 to more than 200 keV/μm.

It is also important to note that the effect of dose modifying agents (including oxygen) diminishes as LET increases, because there is more 'direct' and less 'indirect' radiation injury. At low LET values around 0.3 keV/μm (X and γ-rays) the OER is about 2.5; it falls slowly with increasing LET until about 50 keV/μm above which there is a rapid fall. The OER reaches unity at 300 keV/μm (Fig. 13.6).

The RBE is also dependent on the dose per fraction when comparing radiations of different LET values (Fig. 15.2, Ch. 15). This occurs because of the large shoulder usually present on cell survival curves given with X-rays while high LET radiations are associated with survival curves with little or no shoulder. The loss of shoulder region at high LET is due to a reduction in the capacity for both sub-lethal and potentially lethal damage. As a result low doses of high LET radiations are much more effective than low doses of X-rays for example, but as the size of dose fraction increases, a constant RBE will be reached. The RBE value of fast neutrons measured by large single dose exposures is usually about 2 which significantly underestimate their relative effectiveness when multiple small fractions of dose are given (RBE of 3 or 4). This is the explanation of the 'fractionation trap' which led to many patients in the first neutron therapy trial having severe high dose effects that is discussed in Chapter 15.

It has been shown too that the sensitivity of the cell depends on its position in the cell cycle. This effect is smaller with high LET than with low LET radiations, particularly when low doses of radiation are employed.

The possibilities of differences in repopulation in tumours and normal tissues have also been examined after high LET radiations and in the small number of experiments reported, no differences have been found in the rates of repopulation after high LET or low LET radiations.

Radiation dose rate

Study of cell survival curves will show that as the dose rate is decreased below that normally used for external beam therapy the killing effect is reduced on well-oxygenated cells and the extrapolation number tends to unity (see Ch. 7). At low rates of X and γ-radiation much of the accumulated damage is repaired during irradiation. A difference in response is evident down to about 10 rads per hour when all repairable damage is shed during the irradiation with X or γ-rays. Under hypoxic conditions, however, the dose-rate dependence is much reduced—probably because of greatly reduced recovery from sub-lethal damage during prolonged hypoxia. This is an important biological effect seen with low dose-rate X and γ-radiation. Indeed it will be recognized that the principal effects in radiotherapy of reducing radiation dose-rate are the result of these differences in recovery from sub-lethal damage. (Ch. 14).

In some situations the rate of depopulation by radiation death may be equalled or exceeded by repopulation of the tumour or tissue during irradiation so that differences in cell kinetics may often explain apparent differences in dose-rate effects in different tumours and normal tissues.

Volume factor

It has been long recognized that the clinical response to irradiation is influenced by the volume of tumour and normal tissues contained in the treatment zone. Essentially the response will depend on the degree of cell depletion, but more complex factors may also be involved. These may include damage produced to vascular tissue, persistence of cellular elements to repopulate the area irradiated and any toxic products resulting from cell destruction. The greater the volume of tumour, the greater the dose of radiation will obviously be required to destroy all the tumour cells (Fig. 13.7). This is dealt with in greater detail in the following chapter. The influence of volume on radiation response also depends on the ability of the tissue to contract which is an important process of restitution in skin and in the intestine for example.

In general the larger volume of normal tissue irradiated the greater the reaction to a given dose of radiation. The increase in incidence of radiation nausea with increase in irradiated volume is well recognized. Quantitative data on the volume factor comes from analysis of skin reactions and is described below. Similar data is not available that relates the effects of radiation to the volume of other tissues irradiated such as the brain, spinal cord, kidney, lung and intestine, but the evidence suggests a very similar inverse dose-volume relationship. In considering the sterilization of tumours the volume is critically important as cure depends on the absolute number of tumour cells killed. As the volume of tumour increases so the dose of radiation required to kill all the

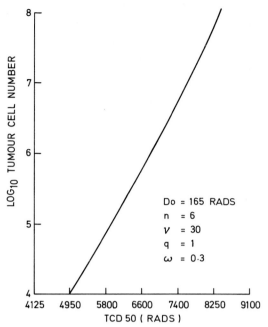

Fig. 13.7 Theoretical relationship between the number of tumour cells and the mean lethal dose (TCD 50) for model tumours.
(From Suit, H. In *Textbook of Radiotherapy.* Edited by Fletcher, Lea & Febiger, 1973).

cells increases. In the following chapter, which deals with clinical application of these principles, the implications of the size of the treated volume are considered as it affects the *therapeutic ratio.* (Ch. 14)

Dose-time relationships

A number of other physical factors are of major importance in determining the biological response. They describe the manner in which the total dose of radiation is determined and it is useful at this point to define the terms used to record these factors.

The term *fractionation* is sometimes used to describe the full details of a course of radiotherapy, but strictly, it should define only the *number* of treatment sessions. It is also necessary to state the *frequency* of treatment sessions which is the time interval between fractions given to each field. It is also extremely important to describe if all treatment fields are treated daily or, as sometimes happens, on alternate days; the biological effect is significantly different. The *dose per fraction* is normally the same throughout the course of treatment, but any major change in the dose given at each fraction will greatly influence the proportion of cells killed.

The *overall treatment time* is also of great influence on the end result of a course of radiation and to increase the overall time is known as *prolongation.* Most courses of radiotherapy are given as a single uninterrupted treatment, but some clinicians advise 'split-course' regimes. It is always essential to describe the *periodicity* of a treatment to indicate the time interval of these planned interruptions to treatment.

The influence of *dose-rate* on the biological effects of radiation has already been described. The practice of delivering the radiation at lower dose-rates is known as *protraction*. The indications for the current clinical re-evaluation of the technique are discussed in Chapter 14.

Chronobiological effects

The timing of fractionation schedules may also prove inappropriate if circadian rhythms are ignored. Figure 13.8 illustrates the consequence of this so-called chronobiological effect on the radiosensitivity of mouse intestine. The

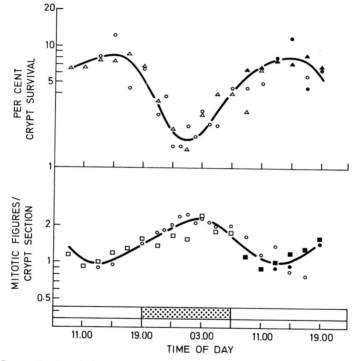

Fig. 13.8 Curves showing whole crypt survival after 1300 rads single exposure and the mean mitotic figure per crypt section.
(From Hendry, 1975).

upper part of the figure shows the considerable variation in crypt survival (assayed by the technique described in Ch. 6, Fig. 6.4) if the mice receive the single dose of 1300 rads γ-irradiation at different times of the day. The lower part of the figure shows that the mitotic index is maximal during the night and that this is when the intestine is most radioresponsive in such nocturnal animals. There is evidence of a similar variation in the radiation response of bone marrow.

Since circadian rhythms are a feature of most biological systems it follows that it may prove to be more beneficial *not* to treat human tumours at the time of peak mitotic activity of limiting normal tissues such as gut and bone marrow. This would mean avoiding radiotherapy during the hours of the day when it is normally administered; but only if the circadian rhythm of the tumour cell population is shown to be significantly different from that of the normal tissues. Only then would the reduction in gut and bone-barrow damage justify such a change in the practice of radiotherapy.

Iso-effect curves and fractionation formulae

Much clinical information has been collected about dose-time relationships and to a lesser extent dose-time — fraction relationships. Unfortunately most of the data is non-uniform and cannot reasonably be correlated. Still less is reliably known about the influence of volume on the biological response during a course of radiotherapy. The relationships of these factors may be described in an equation; solutions of which when expressed graphically become known as iso-effect lines.

$$D_N = D_1 . T^P . N^r . V^q$$

D_N is the dose given in N treatment fractions
D_1 is the biologically equivalent dose given in a single fraction
T is the overall treatment time
P is the function related principally to cell proliferation
N is the number of fractions
r is the function related to recovery from sublethal radiation damage
V is the volume of the irradiated zone
q is the function related to field size or volume.

The qualitative importance of the volume factor has been described above. From studies of skin reactions the volume exponent that has a negative magnitude (increasing volume requiring less dose for the same effect) has been estimated to be —0.33 ($D \alpha V^{-0.33}$). Caution obviously must be advised in applying this exponent to other tissues because of the inadequate data available.

The remainder of the discussion will consider what are called 'dose-time' relationships in which the volume factor may be ignored.

Most information pertains to normal tissues, particularly the response of skin, but some data have been reported which give iso-effect lines for several human tumours (Friedman and Pearlman, 1955); Hall and Howes, 1947); Scott and Brizel, 1974).

The classical study is that reported by Strandqvist in 1944 describing dose-time relationships for certain levels of skin damage (erythema and tolerance) and for the cure of skin cancers. (Fig. 13.9). It was claimed that the dose-dependent responses gave a straight line when plotted on a log-log graph and that the slope of these lines for all degrees of skin damage and for the cure of skin cancer were the same. The slope was increased as a function equal to 0.33

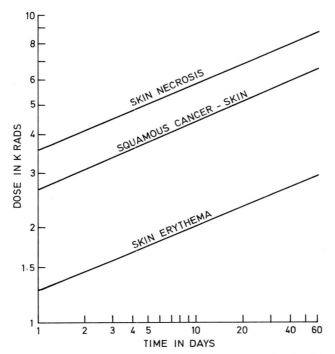

Fig. 13.9 Iso-effect curves relating the total dose and overall treatment time for skin necrosis, cure of squamous carcinoma and erythema.
(After Strandqvist, 1944).

and so it was concluded that for the same biological result dose should vary proportionally to time $(T^{0.33})$ to that power. When only the overall treatment time changed:

$$D \propto T^{0.33}$$
$$D_N = D_1 \cdot T^{0.33}$$

Later studies by Cohen have given a less rapidly varying function for the cure of skin cancer and it is thought that the current data indicates a function of 0.22 for skin cancer. It is to be pointed out that much of this mathematical description of tissue responses is based on rather slender clinical data often without truly comparable control cases. There is still a need for further carefully controlled clinical studies to be conducted to provide adequate and reliable data.

Ellis (1965) has studied in much detail clinical information on high radiation dose levels that were considered to produce a small and acceptable proportion of high dose effects on normal tissues. Based on this data he has proposed an iso-effect formula which, for *normal tissue tolerance*, considers the number of fractions and the overall treatment time. Ellis related the physical dose given in a prescribed overall time in a specific number of fractions to what he called the nominal standard dose or NSD The NSD unit of 'iso-effective' dose has been called the *ret* (rad equivalent therapy).

$$D_N = NSD \cdot T^{0.11} \cdot N^{0.24}$$

These functions are consistent with the slope of the iso-effect times which provided the basis for Strandqvist's studies, but separates the effects of time and the number of fractions. Fowler and Stern (1963) had first made this important distinction based on the results of experiments on pig skin and Ellis found that their data correlated with his clinical experience.

The use of the Ellis formula has provided a means for radiotherapists to compare the *tolerance levels* of doses delivered by different treatment regimes in respect of this effect on normal connective tissues. It should be remembered that it does not represent the iso-biological effect on all tissues or at all dose levels and does not, in itself, indicate the relative effectiveness of treatment techniques. Ellis has pointed out that it is meant to apply to connective tissue injury and so may be relevant to tissues other than skin, and he did include data on other organs such as the kidney. The Ellis formulation should not be used for less than 5 or for more than 30 fractions and the term ($T^{0.11}$) does not apply beyond 100 days.

The concept of NSD may readily be misused in other ways. Commonly for example, the calculated NSD for incomplete treatments (i.e. sub-tolerance levels) are added together although it is inherent in the NSD concept that it is applicable only to regimes of treatment delivering full tolerance doses of radiation. In order to simplify the system and make it more generally applicable Orton and Ellis (1973) introduced factors which relate time, dose and fractionation and which are proportional to *partial tolerance.* These TDF factors provide a simple and more reliable way of comparing and equating *all* regimes of radiotherapy.

It has been recognized that the Ellis formulation does not represent all the physical factors which may influence the biological response. One of the most highly developed mathematical descriptions of the effects of radiotherapy is that reported by Kirk, Gray and Watson (1971). They believe that Ellis' formula is relevant to all degrees of biological effect on normal tissues and not just meaningful in relation to tolerance levels of radiation damage. If this is accepted the NSD notation may be replaced by a parameter which is a constant of proportionality (R)

$$\text{i.e. } D_N = R \cdot T^{0.11} \cdot N^{0.24}$$

Since R for fractionated treatments is simply a measure of the accumulated biological response, they have suggested that R becomes known as the *Cumulative Radiation Effect* or CRT a scale of biological effect. The NSD is therefore the value of CRE at the tolerance level of normal tissue radiation reactions. The difference of this approach to that of Ellis is illustrated by developing the equation given above to describe R or the CRE.

$$R = D_N \cdot T^{-0.11} \cdot N^{-0.24}$$

The following terms are introduced as follows:

R = CRE

d = dose per fraction $\dfrac{DN}{N}$

x = ratio of total time to number of fractions $\dfrac{T}{N}$

CRE = $dN \cdot x^{-0.11} N^{-0.11} \cdot N^{-0.24}$

    = $d \cdot x^{-0.11} \cdot N^{0.65}$

It can now be seen that the CRE is proportional to the total dose, whereas the NSD was chosen to be dependent on the number of fractions. The scale of radiation effect is, therefore, different in the CRE formulation although the effects are the same at tolerance levels of damage.

The concept of CRE has been applied to treatments by moulds and implants and extended to incorporate the influence of the volume factor. It is also possible for the analysis to deal with gaps which may be introduced into a radiotherapy fractionation scheme either inadvertently, or by design, to complete a comprehensive system which may allow the assessment and comparison of the effect of different treatments on normal tissues allowing for a delay in CRE in intervals between treatments.

It must be stressed that the validity of the hypothesis of iso-biological effect on normal tissues in response to different radiation treatment regimes has still to be tested and confirmed. A particular danger arises in using factors derived from immediate reactions and assuming that they may be applied to late effects. The clearest example of this is in excessive prolongation that can avoid or greatly reduce immediate reactions, but not the late effects of radiation.

It certainly has to be appreciated that even if it is shown that treatment regimes may have iso-biological effects on normal tissues that these treatments may not necessarily be *iso-effective* in respect of cancer control. Many schemes of fractionated radiotherapy have been developed empirically that are satisfactory in relation to normal tissue tolerance, but it has yet to be proved that any particular regime is optimal with respect to the maximum local control of cancer and minimum associated morbidity.

The challenge to clinicians and radiobiologists is to develop more precise iso-effect formulae that not only will allow more exact and detailed comparisons to be made of different radiotherapy regimes, but also will provide the basis for the determination of optimal fractionation schedules.

REFERENCES

Cohen, L. & Shapiro, M. P. (1952) Radiotherapy in breast cancer. *British Journal of Radiology*, **25**, 636–645.
Ellis, F. (1965) The relationship of biological effect to dose-time fractionation factors in radiotherapy. Elbert, M., & Howard, A. Eds. *Current Topics in Radiation Research*, **4**, Ch. 7. pp. 357–397 Amsterdam: North Holland.
Fowler, J. F. & Stern, B. E. (1963) Dose-time relationships in radiotherapy and the validity of the survival curve models. *British Journal of Radiology*, **36**, 163–168.
Friedman, N. & Pearlman, A. W. (1955) Time-dose relationship in irradiation on recurrent cancer of the breast: iso-effect curve and tumour lethal dose. *American Journal Roentgenology, Radium Therapy and Nuclear Medicine*. **73**, 986–998.
Hale, C. H. & Homes, G. W. (1941) Carcinoma of skin. Influence of dosage on success of treatment. *Radiology*, **48**, 563–569.
Hendry, J. H. (1975) Diurnal variations in radio-sensitivity of mouse intestines. *British Journal of Radiology*, **48**, 312–314.
Hewitt, H. B., Chan, D. P. S. & Blake, E. R. (1967) Survival curves for clonogenic cells of a murine keratinising squamous carcinoma irradiated in vivo and under hypoxic conditions. *International Journal of Radiation Biology and Related Studies in Physics, Chemistry and Medicine*, **12**, 535–550.
Kirk, J., Gray, W. M. & Watson, E. R. (1971) Cumulative radiation effect. Part I. Fractionated treatment regimes. *Clinical Radiology*, **22**, 145-155.

Orton, C. G. & Ellis, F. (1973) A simplification of the use of the NSD concept in practical radiotherapy. *British Journal of Radiology,* **46,** 529–537.

Regaud, C. (1922) Influence de la durée de radiation sur les effects determines dans le testicle par le radium. *Comptes rendus des seances de la Société de biologie et de ses filiales,* **86,** 878–890.

Scott, R. M. & Brizel, H. E. (1974) Time dose relationships in Hodgkin's Disease. *Radiology,* **82,** 1043–1048.

Thomlinson, R. H. (1967) Oxygen Therapy—Biological considerations. In Modern Trends in Radiotherapy p.52. Eds. Deeley, T. J. & Wood, C. A. P. London: Butterworths.

Widman, B. P. (1941) Radiation Therapy in cancer of the skin. *American Journal of Roentgenology, Radium Therapy and Nuclear Medicine.* **45,** 382–394.

Withers, H. R. (1975) Cell cycle distribution as a factor in multifraction irradiation. *Radiology,* **114,** 189–198.

FURTHER READING

The relationship of time and dose in the radiation therapy of cancer. *Frontiers of Radiation Therapy and Oncology.* **3.** Ed., by J. M. Vaeth. S. Karger (1968). Baltimore: Basel & Upp.

14. Clinical Applications

In clinical practice the radiotherapist is commonly faced with the dilemma of attempting to ensure a high probability of eradicating the cancer within the irradiated volume while at the same time is prepared to accept only a low probability of producing severe late normal tissue damage. The ratio of these two probabilities may be regarded as a measure of the *therapeutic ratio* of the radiation technique.

The relationship of probabilities for tumour sterilization and for the induction of serious late effects in normal tissues at different sites for the wide variety of treatment regimes is poorly documented and much detailed clinical research remains to be undertaken in this field. A number of animal experimental studies (Suit, 1972) have elegantly illustrated the methods by which this relationship may be evaluated and clinical experience (Fletcher & Shukovsky, 1975) also supports the application of these concepts to the practice of radiotherapy.

Dose-reponse curves

The principles of fractionation given in Chapter 13 were described in terms of cell-survival curve parameters. The effects of different regimes of radiotherapy on tumours and normal tissues may be described quantitatively in terms of dose-response curves. Dose-response curves essentially are related to the relevant cell-survival curves and depend on the initial number of clonogenic cells in the tumour or normal tissue population and the fraction of surviving cells. The tumour control probability, that is the number of tumours sterilized compared to the number irradiated, may with certain assumptions be given by the exponent $(-S N)$. N represents the number of original cells in the tumour and S the surviving fraction which in turn is dependent on the dose of radiation delivered and the cell survival curve parameter, Do. In this way the quantitative cell-survival data may be correlated with the quantitative responses of tumours and normal tissues that are observed in clinical applications.

Figure 14.1 illustrates the percentage cures of a transplanted C3H mouse mammary carcinoma and the incidence of necrosis in the overlying skin following three different treatment schedules with X-rays. Although data are relatively incomplete in respect of human experience, sufficient information on the clinical dose-response of tumours and normal tissues is available to confirm that they are similar to those obtained from animal experiments. There are, of course, important quantitative differences in the responses measured in different tumours in various species. For example, a single X-ray dose of about 5000 rads is required to cure the mammary carcinoma in the mouse, but we know that a single dose of 2000 rads is usually sufficient to cure human skin cancer of apparently similar size.

The typical dose-response curve is sigmoid in shape (Fig. 14.1) and a large family of similar curves may be obtained by varying the biological or physical factors that were described in Chapter 12. It will be seen that these curves are similar in shape to those obtained from whole-body exposure of animals (Fig. 9.1), whose death is simply a reflection of the depletion of stem cells below a critical level in essential tissues, such as the brain, bone-marrow or gastro-intestinal tract.

It is important also to note that the dose-response curves for tumours and normal tissues are apparently indistinguishable in their shape and slope. The relative radio-responsiveness of a tumour or normal tissue may be measured by their position on the dose scale and, all other physical and biological factors being constant, tumours or tissues may be regarded as increasingly radio-responsive as the position of their response curve moves to the left. The curves of tumours and normal tissues are similar because they both depend predominantly on two biological factors, firstly the histological type of the tissue or tumour and secondly the volume (or number) of normal tissue or tumour cells that are within the fields of irradiation.

Several principles of particular relevance to clinical radiotherapy may be illustrated from more detailed consideration of these dose-response curves. The first and most important point to be made is that the slope of these dose-response curves is extremely steep and this has also been confirmed in human tumours (Fig. 14.2). This means that a *small* increase in radiation dose will produce a *large* increase in local cure rates. Likewise a small decrease in the dose will result in considerable increase in local recurrences of tumour. This point emphasizes the critical importance of adhering to a optimum dose and treatment schedule once it is decided. It may be possible in consideration of the relationship of the tumour and normal tissue curves to increase the probability of cure in certain cancers by an increase in dose that may produce a comparatively small increase in the probability of normal tissue damage. However, a point will be reached at which increased dose will result in a much greater incidence of severe normal tissue damage than is acceptable. The essential importance of dose in respect of both local tumour control and normal tissue damage emphasizes the importance of accurate beam direction and homogenous dose distribution in the high-dose volume in good radiation treatment planning. Inhomogeneity of dose will prejudice both the probability of tumour control and the avoidance of serious morbidity in normal tissues.

The term *tumour lethal dose* was at one time used to describe the lowest radiation dose that would be effective in destroying tumours of a particular histological type and size. It is obvious from Figure 14.1 that after a threshold dose there is at first a slow increase in probability of local tumour control followed by a steep increase in cure probability. Strictly, there is no single tumour lethal dose that may be quoted for each histological type and size of tumour. The greater the dose of radiation delivered the greater the probability of tumour sterilization. It is reasonable, however, to attempt to define a *mean*

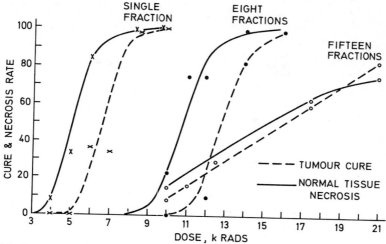

Fig. 14.1 Dose response curves for the cure of the C3H Mammary carcinoma and necrosis of overlying skin for three different schedules of fractionation.

tumour lethal dose, that would sterilize 50 per cent of tumours of the same type and size using a certain radiation treatment regime. The mean lethal dose is a value that may be regarded as a measure of the radio-sensitivity of tumours in terms of its cellular response and is a valid parameter for comparing the

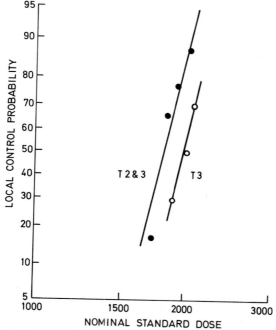

Fig. 14.2 Dose-response curves for human cancers of the oropharynx (T_2 & T_3 cases) from Shukovsky & Fletcher (1973) and for laryngeal cancers (T_3 cases) from Stewart (personal communication).

effectiveness of radiation in sterilising different tumours. Such values, although available for many experimental tumours are lacking for human cancers.

Another important principle that may be established from further examination of dose-response curves (Fig. 14.1 and 14.2), is that there is no evidence that radiation doses above that associated with 100 per cent cure will result in a decrease in tumour control probability. It was at one time suggested that excessive radiation dose did increase the risk of recurrence, but this is not borne out by experimental observation. The concept was called the *supra-lethal effect*, but it is not a real phenomenon. It seems that the hypothesis may have been based on the occasional clinical observation that unresponsive tumours may remain in an area irradiated to high-dose levels, and in which a severe immediate radiation reaction had been produced. It was thought that the radiotherapy had caused immediate necrosis of normal tissue and yet allowed tumour to survive. We now know that many biological factors, but particularly the kinetics of the irradiated cell populations contribute to this differential effect and one need not evoke an anomalous dose-response relationship. The association of tumour recurrence and normal tissue necrosis will be found regularly on the grounds of mathematical probability, but particularly often in clinical situations where the therapeutic ratio is low.

The dose-response curves (Fig. 14.3) also illustrate the constant dilema that faces radiotherapists. As the volume of cancer enlarges by an increase in the

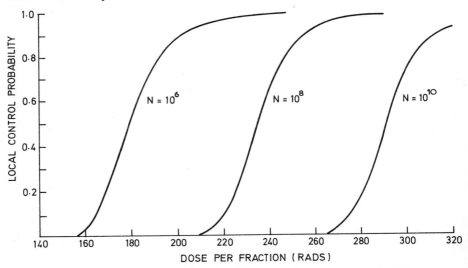

Fig. 14.3 Calculated dose-response curves for different numbers of cells in a tumour.

number of cancer cells, so the dose of radiation required for a certain probability of local eradication of the tumour is increased. However, as the volume of related normal tissues will also be increased, the dose of radiation for the accepted level of morbidity, will have to be reduced. The probability of cure in radiation therapy is always determined by the acceptable level of morbidity to normal tissues that must necessarily be included within the high-dose volume.

The radiation dose that will produce the acceptable level of morbidity is known as the *tolerance dose* and a detailed knowledge of tolerance doses for different normal tissues, is essential for the safe practice of radiotherapy.

It will be seen that the dose-response curves for larger tumours are moved to the right, indicating that a larger radiation dose is necessary to maintain the same probability of local control. As the size of the tumour increases so the mean lethal dose increases. At the same time the curve representing the incidence of late tissue effects will approach the tumour response curve as the volume of tissue irradiated increases. Accordingly the margin of safety is much reduced and it will be found in these circumstances that the *therapeutic ratio* becomes so low that curative or radical radiotherapy may no longer be feasible. It has been found in practice that fractionation often helps to improve the effectiveness of radiotherapy, especially when large tumours have to be treated and this is illustrated for skin tumours in Figure 14.4.

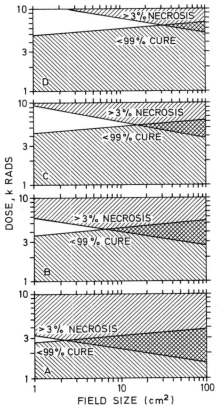

Fig. 14.4   Display of relationship of cure (90 per cent and necrosis (3 per cent) for skin cancer for different schedules of fractionation; A. 1 day, B. 5 days, C. 21 days, D. 45 days. (From Von Essen, 1963).

A policy of *radical radiotherapy* may therefore be defined as treatment designed to provide a high probability of local tumour control while accepting the probability of a small (but not zero) number of serious normal tissue effects.

Radiotherapy regimes that produce the maximum acceptable level of normal tissue morbidity are known as *tolerance doses* of radiation. Tolerance doses require to be defined for each different tissue, quality of radiation, volume irradiated, number of fractions and overall time. The incidence of morbidity that is acceptable depends on the nature and site of the radiation effect and also the possibility of its successful repair. The definition and clinical significance of normal tissue tolerance is discussed below.

Palliative radiotherapy has to be offered when a satisfactory therapeutic ratio cannot be obtained for the local eradication of a primary tumour, or when the cancer is so widely disseminated that cure of the widespread disease is not possible. The aim of palliative treatment is to produce quickly a significant tumour response in the treated area and worthwhile restraint of its growth with little or no related morbidity.

Normal tissue tolerance

Normal tissue tolerance is the expression used to define the dose of radiation which produces the maximum acceptable number of serious late changes. The character of these late effects depends on the tissue concerned, but the exact nature of the radiation effect should always be carefully defined in describing morbidity after radiotherapy as the severity of lesions varies greatly. They may relate simply to severe atrophy and telangiectasia of the skin, or more seriously to frank necrosis or ulceration. In the gastro-intestinal tract it may amount to foci of superficial ulceration, or to gross stenosing enteropathy. The nature and clinical significance of the effect will therefore vary from tissue to tissue and from site to site, as described in Chapter 11.

The use of high doses of radiation that must carry some risk of serious mobidity is important in the treatment of many common cancers which are of limited radio-responsiveness. The degree of risk of radiation necrosis of normal tissues which experienced clinicians may be prepared to accept is determined by a number of considerations. The most important of these is the clinical significance of the particular high-dose effect. If the site of radiation necrosis is a vital centre such as the mid-brain, the lesion will be fatal and is clearly unacceptable. Similarly, irreversible damage to the spinal cord that would produce a paraplegia is not a justifiable hazard of radical radiotherapy. Soft tissue necrosis of skin or gastro-intestinal tract that is remedial by conservative measures or surgical repair is, however, acceptable in order to achieve high cure rates, provided the incidence of high-dose effects is not too great and the associated mortality low. Most radiotherapists would be prepared to accept levels of 3 to 5 per cent necrosis in many sites and, in the treatment of advanced laryngeal cancer a necrosis rate of 10 per cent is often the limit of acceptable morbidity when laryngectomy is feasible.

The tolerance dose of normal tissues is not an absolute value. Tolerance doses, of course, differ for each tissue and depend on the biological factors mentioned above and described in some detail in Chapter 12. The tolerance dose of skin, for example, varies not only for different schemes of fractionation

(Fig. 14.4), but also with volume and site irradiated; tolerance doses for the skin of the trunk are much less than for the skin of the face, for example.

It should now be clear that to achieve maximum tumour control rates the largest possible dose of radiation that may be tolerated by normal tissues may have to be given.

Therapeutic ratio

The *therapeutic ratio* of a radiotherapy regime may be considered as the ratio of the probability of local cancer control compared to the chance of producing serious late normal tissue effects (Fig. 14.5). The concept is illustrated by relating the tumour response to serious normal tissue effects for a wide range of

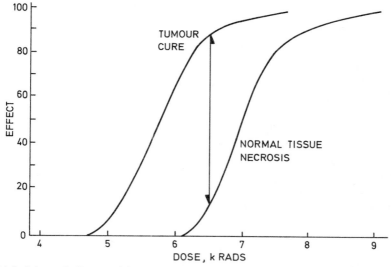

Fig. 14.5 Schematic diagram of dose-response curves for tumour cure and normal tissue necrosis.

dose levels from the data of the dose-response curves for both tumour and normal tissues (Fig. 14.6). Should the sigmoid dose-effect curve for tumour and normal tissues coincide, the resulting curve of therapeutic ratio will be a straight line; a dose giving a certain probability of cure will carry an equal probability of producing normal tissue necrosis. In clinical practice the response curve of normal tissues must always lie to the right of the tumour response curve in order to obtain favourable therapeutic ratio. Initially it may be seen that for a rapidly increasing probability of cure little increase in normal tissue morbidity is observed. A point is reached when the relationship changes, and the probability of tumour cure is increased only at the expense of a disproportionate and unacceptable increase in the risk of normal tissue damage. This point equates with *optimum dose* for that particular type and size of tumour at a specific site and using that particular radiotherapy regime. It is possible to consider these concepts in more detail by analysing the results quantitatively using the pair of related tumour and normal tissue dose-

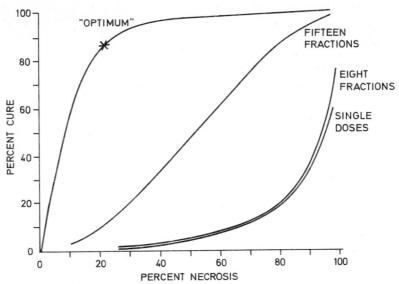

Fig. 14.6   Relationship of tumour cure to normal tissue necrosis following irradiation of the C3H mammary cancer described in Figure 14.1. The normal relationship in clinical radiotherapy is illustrated by the hatched curve.

response curves (Fig. 14.6). At the optimum dose in this particular example, 85 per cent of the cancers will be locally controlled at the expense of 10 per cent soft tissue necrosis. One may assume that the cases of necrosis may occur proportionally between the 85 per cent cured patients and the 15 per cent with recurrences. At the selected optimum dose level of 6500 rads for this particular group of cancers it can, therefore, be estimated that 1.5 per cent of cases will suffer soft tissue necrosis, but will have no evidence of recurrent cancer. 76.5 per cent of cases (85 per cent minus 8.5 per cent) will have uncomplicated local control of the cancer while about 1.5 per cent of cases will have both persistence of the tumour and necrosis; 13.5 per cent (15 per cent minus 1.5 per cent) will have recurrent cancer in the treated area. In clinical practice it may often be found that the incidence of recurrence combined with necrosis may be slightly higher than expected. Secondary factors, such as infection associated with recurrence or the trauma of repeated attempts to confirm recurrence by biopsy, may increase the probability of a breakdown of highly irradiated normal tissues and so influence the actual incidence rates observed clinically.

It has to be realized that a whole family of curves exist that describe the clinical relationship of cure and morbidity. It is possible by using improved fractionation techniques (certainly in the treatment of experimental animal tumours) or by other methods described in Chapter 15 to increase the therapeutic ratio and enhance the effectiveness of radiotherapy in cancer control. Figure 14.7 illustrates the quite remarkable improvement in therapeutic ratio that may be obtained with different schemes of fractionation and by the use of fast neutrons instead of X-rays in the treatment of a transplanted C_3H mammary tumour. The radiobiological data necessary to design rational and

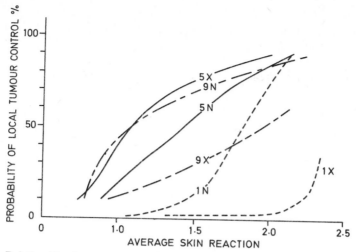

Fig. 14.7 Relationship of tumour cure to immediate skin reaction in the treatment of the mouse mammary carcinoma by X-rays and fast neutrons using 1, 5 or 9 fractions. (From Fowler *et al.*, 1972).

optimum treatment regimes in patients is not yet available. Much painstaking clinical research remains to be undertaken to determine the factors that will favourably influence the effectiveness of radiation in sterilizing tumours and to evaluate new techniques that may improve the therapeutic ratio.

Radio-sensitivity and radio-curability

The term *radio-sensitivity* is normally applied to the quantitative assessment of the depletion of clonogenic cells following irradiation and it is advised that reference to 'radio-sensitivity' be related to cellular effects.

When the clinical reaction of tissues or organs to radiation is considered it is more appropriate to describe the biological effect in terms of *radio-responsiveness*. The complex radiation response in tumours and normal tissues depends not only on the radio-sensitivity of the individual cells, but also, as far as the clinical effect is concerned, on the cell proliferation kinetics of the tissue, the effect on the vasculo-connective tissues and the functional integrity of homeostatic repair processes. Tissues with a rapid turnover of cells will be much more radio-responsive than slowly proliferating cell populations, although the intrinsic cellular radio-sensitivity of the tissues may be similar in both cases. Thus the response of the bone marrow, a rapidly proliferating cell system, is manifest by a rapid fall in the number of cells following a dose of radiation. Irradiation of the liver on the other hand, that is composed of a slowly proliferating population of cells, is followed by the slow depletion of damaged cells with little apparent radiation response to be seen immediately after irradiation. It has been pointed out in Chapter 6 that the cellular radio-sensitivity of all mammalian cells is very similar and so in the example of the bone marrow and the liver, the number of cells killed by a single large dose of radiation will be virtually the same, although the more rapidly dividing tissue in bone marrow will show evidence of loss of tissue cells more quickly. Strictly, it is differences in cell population kinetics that influences the radio-responsiveness of normal tissues and tumours and is the basis of the Law of Bergonie and Tribondeau described in Chapter 3.

It has been made clear that although the observed differences in radio-responsiveness of different tissues and tumours may be great the absolute numbers of cells suffering lethal radiation damage may be little different. Since the probability of tumour cure after radiation is related to the initial number of clonogenic cells, it will be understood that the radio-responsiveness of tumours cannot be correlated with *radio-curability*. Certain cancers, such as the slow-growing cystic basal cell carcinoma and the adenoid cystic carcinoma, may frequently take many months to resolve following a curative course of radiotherapy. In time, complete and permanent regression may be obtained. By comparison, a highly undifferentiated rapidly growing cancer may disappear quickly, sometimes during the course of radiotherapy, but if it is not cured may recur only a month or two later. The degree of resolution of a tumour at the end of a course of radiotherapy may not, as a general rule, be taken as an index of its radio-curability since the speed of regression of the tumour is independent of the degree of tumour cell sterilization. However, much more data is required relating the speed of resolution of different types of tumours with the probability of local irradication (Suit *et al.*, 1966). It should be accepted at present that a pre-planned course of radiotherapy should not be reduced even if the tumour is apparently responding well during treatment.

It has been shown in Chapter 4 that it is the processes of cell replication that are sensitive to the effects of radiation and that there is no great variation in radiosensitivity between different kinds of cells. It will now be understood that it is necessary to revise some traditional dogma about the so-called radio-resistance of many cancers for the assessment has often not been well-founded. Some examples may be given of the misconceptions that may arise from the wrong interpretation of clinical observations.

Many salivary tumours are characteristically slow growing and may reach very large size before presentation; this is true also of some soft tissue sarcomas. The failure of such large slow-growing tumours to respond completely to a course of radiation is often primarily a reflection of their size and the impossibility of delivering safely to such a large volume a radiation dose sufficient to sterilize all the tumour cells. The slow growth characteristics of these tumours also tends to mask the effectiveness of the radiotherapy in terms of cell killing. In other situations it is the highly radio-responsive nature of the tissue of origin, such as the small intestine or colon that may make the adeno-carcinoma of the bowel appear by comparison highly radio-resistant. The immediate response to radiation certainly is much greater in the normal intestine, but it is wrong to assume that adeno-carcinoma of the gastro-intestinal tract is wholly unresponsive to radiotherapy. Oat-cell carcinoma of the bronchus is recognized as highly radio-responsive because of its fast growing nature, high cell turnover and rapid expression of the lethal effects of radiation in its cell population. Tumours that are slow growing much more slowly manifest radiation injury and regression.

The oxygenation of tumours may also influence their responsiveness to radiotherapy. It has been claimed that the scirrhous carcinoma of breast, which is commonly atrophic and poorly vascularised, normally responds less well than its metastases in the regional lymphatic nodes that are considered to have a better blood supply. In contra-distinction a squamous carcinoma in the tongue that is normally well vascularized will often respond much better than associated metastases in lymph nodes in the neck, that may be so large that they have outgrown an adequate blood supply before presentation.

It would seem, therefore, that clinical experience is not at variance with the data obtained from cell survival curves that there is no great difference in the intrinsic radio-sensitivity of different cell types. Differences in the radio-responsiveness and cure rates seen following radiotherapy are largely explained by differences in the size of tumours, critically the number of cancer cells and differences in their cell population kinetics. Tumours of the lymphoid cells and other primary tumours containing large numbers of lymphocytes (such as seminoma and lympho-epitheliomas) will regress more quickly following radiation as the irradiated lymphoid cells will die in inter-phase, rather than await progress to mitosis before dying, as is the case for most other mammalian cells.

Significance of apparently residual tumour on biopsy after radiotherapy

It should now also be understood that after a course of radiotherapy apparently *residual* cancer in the treated area may not be capable of unlimited proliferation and progressive resolution may take place over some months. This has earlier been described in the case of the cystic basal cell carcinoma, that characteristically regresses slowly after radiotherapy. Biopsy during the period of regression will demonstrate cancer cells that are morphologically intact, but if the radiation dose has been adequate, there will be no viable, clonogenic cells in the tumour (Suit and Gallagher, 1964). Considerable experience has to be exercised in deciding that an irradiated tumour is recurrent and essentially the decision has to be made on clinical grounds, on the basis of renewed tumour growth.

Fractionation schedules

The most common schedules of fractionation employ daily doses of radiation with a break on Saturday and Sunday. The optimum doses for individual fractionation schedules have been determined by years of clinical experience and when these regimes are at the limits of tolerance radiotherapists are very naturally unwilling to change any of the factors involved. As a result most radiotherapy departments maintain a policy of using a standard number of fractions and overall time, although it is unlikely that a single fractionation schedule is optimum for every clinical situation. Certainly we do not have the necessary biological data on which to design individual treatments, but there are reports of clinical trials which suggest that daily fractionation may not be necessary to obtain the best results.

The British Institute of Radiology initiated a trial in the United Kingdom that compared the use of three and five fractions weekly in the treatment of larynx and pharynx cancer (Wiernik, G. *et al.,* 1972). The doses given in both the three and five-weekly fraction schedules were chosen to be equivalent, and no difference was anticipated between the two groups. In the report of this prospective random controlled trial, 334 patients were analyzed, 159 patients treated with three fractions per week, 174 with five fractions per week, and no difference was evident either in local tumour control, the severity or number of radiation reactions or survival rates.

Bates (1975) has reported on another prospective clinical trial in which three fractions each week for four weeks was compared with only two fractions each week for three weeks in the post-operative radiotherapy of breast cancer. In this very detailed study 203 patients received 12 fractions and 208, six fractions. The radiation dose, of course, was suitably reduced in the schedule of six fractions, but to an extent that was 10 per cent lower than would have been indicated using the Ellis formula. The local recurrence rate and the morbidity were similar in both groups (Tables 14.1 and 14.2). Bates does stress that the

Table 14.1   Actuarial assessment up to 5 years of results of radiotherapy in breast cancer using two fractionation schedules. (From Bates, 1975).

Overall Time	Three weeks	Four weeks
Fractionation	2 per week	3 per week
Number of patients	208	203
Local recurrence	22 (10.6%)	17 (8.4%)
Distant metastases	44 (21.2%)	42 (20.7%)
Survival	158 (76%)	154 (76%)

Table 14.2   Severe morbidity at one year following two fractionation schedules used in treating breast cancer. (From Bates, 1975).

Overall time	Three weeks	Four weeks
Fractionation	2 per week	3 per week
Number of patients	169	165
Skin stigmata	2 (1.2%)	2 (1.2%)
Pulmonary Fibrosis	18 (11%)	28 (17%)
Lymphoedema	39 (23%)	34 (23%)

dose delivered is more critical with the six fraction treatment, that all fields must be treated on each occasion and the overall time must never be reduced by even one day. Bates' opinions are based on over 15 years' experience and suggest that the use of a large number of fractions, and perhaps daily fractions, may provide a greater margin of safety in the application of radical radiotherapy techniques that deliver doses at the limits of normal tissue tolerance.

At the other end of the spectrum consideration is being given to the application of two or more fractions each day. (Choi & Suit, 1975). This concept is based on our knowledge of two important radiobiological factors in fractionation (Ch. 13). Firstly, it is known that in most mammalian cells recovery from sub-lethal radiation damage is complete by about four hours. Secondly, where a very rapidly growing tumour arises in a tissue with slowly proliferating characteristics, a short-interval between radiation dose fractions will produce a greater cumulative effect on the cancer than on the related normal tissues. Pilot studies using multiple daily fractions have been employed to treat soft tissue sarcomas. Burkitt's lymphoma, glioblastoma and skin tumours. These techniques, using multiple fractions each day, may lead to improved results in very rapidly proliferating tumours such as in Burkitt's lymphoma. (Norin, *et al.,* 1971).

Another less commonly used method of fractionation is the 'split-dose' regime. In this technique a rest period usually of about two weeks is given between a divided course of irradiation. The biological basis for this technique is primarily that the cell population of normal tissues is restored more quickly than the tumour in the rest interval. There may also be an advantage in that the oxygenation of the residual tumour may improve considerably during the interval, as has been shown in certain circumstances and so its response to the second course of irradiation will be enhanced. Another very important practical advantage is that radiation reactions of the skin and mucous membranes is minimized and this may be much appreciated by patients, although the period of treatment is extended. Clinical studies have been conducted but it has not been established that 'split-dose' regimes offer any significant improvement in survival rates or therapeutic ratio over conventional methods of fractionation.

The application of weekly dose fractions over a period of up to three months may be regarded as an extension of this technique. Weekly fractionated treatment may be of value in treating very slowgrowing tumours that are considered normally unresponsive to radiotherapy. The technique used as a palliative procedure, was introduced by Paterson (1948) and is known as weekly 'growth restraint' and is an effective way of controlling large slow-growing growths.

Treatment of all fields at each application

The convention normally implies that every field in the prescribed plan is treated daily; if this is altered so that, for example, only alternative fields are treated daily, the dose per fraction is increased on each field so that the biological effect is greater in each field. The increase is greater in the overlying normal tissue than in the tumour, except in treatment with charged particles. Morbidity may be signficantly increased and the therapeutic ratio may be greatly reduced by this technique.

The detailed pathogenesis of late normal tissue effects has not been fully determined for all tissues and its has already been pointed out that late radiation effects are influenced by the processes of repair as well as the nature of the primary radiation damage (Ch. 11).These pathological repair processes will obviously vary in different tissues and the type and degree of these reactions (such as fibrosis) will influence the development of late radiation effects on normal tissues.

For these reasons tables of tolerance doses for normal tissues must be regarded with caution, particularly because the physical factors influencing the radiation response may vary significantly (Ch. 13). The factors include the radiation quality, dose rate, dose per fraction, interval between treatments and the fraction number and overall time.

The importance of overall homogeneity of dose distribution across the tumour has already been stressed. It is of equal importance in the treatment of deep-seated tumours that no imbalance of the dose given at each treatment is tolerated across the tumour volume as a result of treating only alternate or selected fields every day. In addition, failure to treat all fields at each session will mean that the overlying normal tissues will receive a relatively higher biological dose, as mentioned above. At present radiotherapists normally prescribe a radiation treatment plan by assessment of the iso-dose distribution. It is possible to consider the relative merits of treatment plans in terms of iso-effect, or iso-survival curves. (Holsti *et al.*, 1971). One may also attempt to estimate the effective biological dose by calculating the values of NSD or CRE (Ch. 12). which would better indicate the effect on the normal tissues of different treatment plans. The principle of calculating iso-effect curves is particularly important when one considers the application of high LET radiations, the biological effects of which, compared with the standard low LET radiation, will vary considerably with the size of the dose fraction (Ch. 15).

The differences in biologically effective doses given by various treatment schedules have been analyzed by Wilson & Hall (1971) in terms of the NSD concept (Ch. 13). The differences may be illustrated by considering a radiation

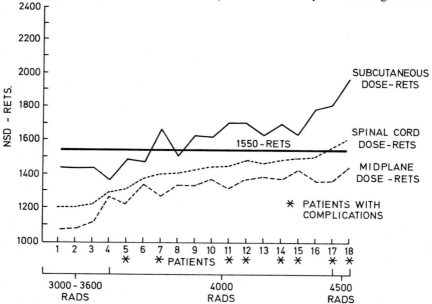

Fig. 14.8 (a) Dose in rets received by patients treated for Hodgkin's Disease indicating the spinal cord dose and those patients who developed evidence of myelitis. Alternate opposed fields were treated daily.

treatment plan using parallel opposing fields. If the inter-field distance (or separation) is 20 cm a central dose of 120 per cent is delivered (60 per cent from each field) and the subcutaneous maximum dose is also 120 per cent (100 per cent + 20 per cent exit dose). It is desired to give 240 rads central dose at each treatment. If both fields are treated each day the 'given' dose from each field would be 200 rads (240 ÷ 1.2). If alternate fields are treated daily the 'given' dose on each field would be 400 rads (240 ÷ 0.6). Assuming the treatment would be given in 20 fractions over 4 weeks the NSD would be 592 rets to the subcutaneous tissues, compared to 1003 rets if treatment was given only on alternate days — producing a much greater damage to the superficial tissues. This effect is much more important when less penetrating beams are being used and when central deep-seated tumours are being treated.

Marks *et al.* (1973) have reported an increased rate of complications in a group of 18 patients with Hodgkin's Disease treated by parallel opposing wide fields, treating only one field daily. Eight patients had serious late radiation effects on normal tissues which have been correlated with excessive doses calculated in terms of the NSD (Fig. 14.8). When the same physical dose is delivered by treating both fields daily, the biological doses are, however, safely below the tolerance levels estimated to be 1550 rets.

Reducing field techniques

It was earlier stated how important it was to ensure homogeneous dose distribution in the treatment volume in order to obtain optimum results in

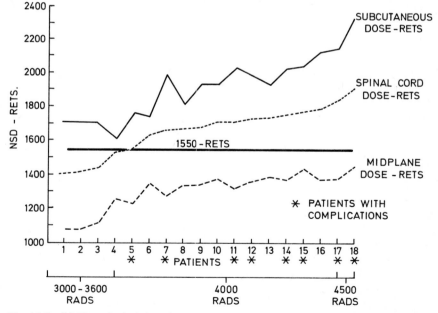

Fig. 14.8   (b) Hypothetical dose in rets for same group of patients assuming both fields were treated daily. The spinal cord dose would then be below the tolerance level of 1550 rets except in patients 17-18 (From Marks *et al.*, 1973).

radiotherapy. This statement does not imply any criticism of those techniques in which the high-dose volume is reduced as treatment progresses and that are designed to deal with two common clinical problems in cancer therapy. Firstly, 'reducing field' techniques allow radiotherapy to be given initially to a large volume of tissue in order to treat small deposits of cancer, for example, in lymph nodes that may not be clinically detectable and represents 'sub-clinical' spread of disease. In this large volume containing perhaps only microscopic spread of cancer a lower dose than is required to control, with the same probability of success, the larger primary tumour may be given, without increasing the risk of serious damage to the normal tissues. Secondly, by reducing the treatment volume during a course of radiotherapy as a highly responsive tumour resolves, the practice of sparing the surrounding normal tissues from the total radiation dose by 'coning down', allows a higher radiation dose to be delivered to the central residual tumour mass without greatly increasing the risk of normal tissue morbidity. However, for optimal effect the radiation dose within the pre-determined high-dose volumes, large and small, should be as homogeneous as possible. In all radiotherapy techniques the regions of high-dose will determine the risk of normal tissue morbidity and regions of low-dose will determine the probability of residual cancer.

Dose-rate considerations

There is no appreciable dose-rate effect in the range of 30 rads/min to 1000 rads/min, that is normally used for external X-ray beam therapy. However, it must be remembered that when equipment with a low radiation output is employed the dose rate at the tumour will be even smaller than the stated output. The dose-rate effect is most clearly discernible between 1 rad/min and 100 rads/min when increasing effectiveness can be demonstrated with increasing dose-rate. Above and below these X-ray dose-rates there is little change in the survival curve parameters (Fig. 14.9). Although one should be aware of possible differences in biological effect with these lower dose-rates, it is unlikely to be of importance, unless large dose fractions are being given. Then it is the long treatment time, perhaps approaching an hour, during which 'recovery' takes place, that is the important issue.

Low dose-rate treatments have been used with surface applicators, interstitial and intra-cavitary techniques. It was considered that radium implants, for example, had some special merit, but this technique has largely been superseded by the introduction of megavoltage external beam therapy. It remains true, of course, that it is possible by implantation, or by intra-cavitary applications to deliver a much higher radiation dose to a limited tissue volume than is possible by external beam techniques. In this way they retain an obvious advantage, since cure probability is dependent on dose. It is also important that at low dose-rates the oxygen enhancement ratio is reduced and these techniques may more effectively deal with anoxic tumours than conventional photon beam therapy.

The effectiveness of these low dose-rate treatments will depend critically on

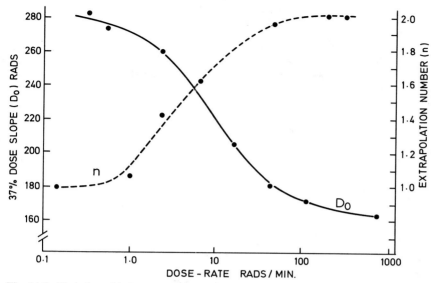

Fig. 14.9 Variation with dose-rate of Do value (continuous curve) and the extrapolation number (hatched curve). (After Hall, 1972).

the dose-rate rather than the total dose of radiation delivered. The effect of dose-rate is illustrated in Figure 14.10. A standard dose for an interstitial treatment has been taken as 6000 rads in seven days. It is found commonly in clinical practice that the dose-rate is increased, either because higher strength radionuclide sources are used, or the geometry of the actual implant produces a smaller treated area than intended. In these circumstances, because of the increased effectiveness of the higher dose-rate a lower total dose must be given to produce the same biological effect as 6000 rads given over 168 hours.

The use of the low-dose rate technique is also being evaluated for external beam therapy (Pierquin, 1970). A dose-rate of 110 to 180 rads per hour from a telecobalt machine has been given in doses of approximately 1000 rads each day for about seven days. It seems from early studies in which patients with advanced cancer of the head and neck region were treated in this way, that doses of the order of 6000 rads may be safely given to parallel opposed fields of 10×10 cm, a very much higher dose than could be tolerated with conventional dose-rates.

The Cathetron was introduced as an important development in improved radiation protection associated with the use of intra-cavitary radio-nuclides. The patient under treatment may be placed in a shielded room and so staff and other patients need not be exposed to radiation. Difficulties immediately arose in calculating the dose of radiation to be given that would be equivalent to the low dose-rate radiation applications. Liversage (1969) has made a detailed study of the physical and biological factors involved and has provided a formula for correlating low dose-rate and high dose-rate regimes. When the protracted treatment time 'T' is greater than twelve hours, the number 'N' of high dose-rate treatments in the same overall time for the same biological effect is given by the formula $N = {}^T/_4$. In practice this would mean that a 96-hour radium insertion delivering 5000 rads at point 'A' could be equated with (96/4) = 24 fractions given at high dose-rate over four days. As six treatments each day would be grossly inconvenient for the patient, the overall time is normally increased to give a single Cathetron treatment perhaps on alternate days and then either the number of fractions or overall time, or both, are corrected using the Ellis NSD equation.

Conclusions

Rationalization of fractionation remains one of the major challenges of clinical radiobiology. Rational fractionation schedules imply a greater certainty of local tumour control associated with an enhanced therapeutic ratio and if reduced fractionation really is as safe and effective as longer regimes, the logistic and economic importance of its general acceptance is obvious.

It is indeed regrettable that in spite of the large volume of clinical experience in radiotherapy, data does not exist on which to base rational, or determine optimum, treatment regimes. Too much attention has been spent in the past collecting information on patient survival rates alone. It cannot be too often, or too strongly, pointed out that for the critical evaluation of radiotherapy it is *local tumour control* and the *morbidity* of normal tissues in the *irradiated volume* that are the valid and essential end-points for comparison and assessment. Survival is undoubtedly important in the final analysis, but survival data may easily give a misleading impression of the real value of radiotherapy. It must always be borne in mind that excessive radiation dosage may produce a decrease in survival of a treated group of patients by increasing the incidence of fatal high-dose effects. This was seen in the management of testicular tumours by radiotherapy before the recognition of radiation nephritis explained an unexpected fall in survival, following the introduction of a new technique. Similarly, in the management of brain tumours, excessive radiation dose may result in fatal brain necrosis which the unwary and uncritical clinician may wrongly assume to be due to failure to control the primary tumour, unless autopsy is performed.

Some examples of the relationship of local control of tumour and local normal tissue morbidity following different radiation doses, or different techniques, are given in Figure 14.11. The results of radium needle implantation of the tongue are of a random controlled trial and give the local control and necrosis ratio at five years. The results of megavoltage irradiation of laryngeal cancer are also assessed at five years, but represent sequential experience and are not obtained from a controlled trial. This is also true for the results of treating early cancer of the penis and in this case two different treatment techniques are compared and the assessment made at two years. They illustrate for squamous cell carcinoma at three different sites the direct relationship between tumour control, necrosis and radiation dose.

Since the application of the principles of clinical radiobiology can only be

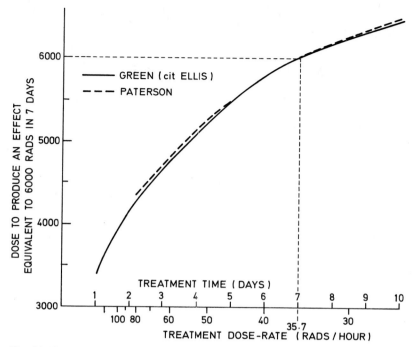

Fig. 14.10 Relationship of total dose to overall treatment time for a radium implant to produce an effect equivalent to 6000 rads in 7 days. (From Hall, 1972).

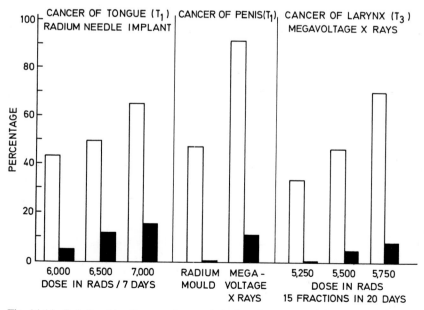

Fig. 14.11 Relationship of cure and necrosis to dose for cancers of the tongue, larynx and penis. The open areas indicate tumour control, the closed areas normal tissue necrosis.

expected to lead to improvements in *local* responses, it is to be hoped that published reports of the results of radiotherapy will in future give detailed accounts of the quantitative assessment of the local effects on tumour and normal tissues. It is only when these essential clinical data are presented, together with precise statements of technique and dose, that a meaningful evaluation may be made of the many differences to be found in the current practice of radiotherapy throughout the world.

REFERENCES

Bates, T. D. (1975) A prospective clinical trial of post-operative radiotherapy delivered in three fractions versus two fractions per week on breast carcinoma. *Clinical Radiology*, 26, 297-304.
Fowler, J. F., Denekamp, J., Page, A. L., Begg, A. C., Field, S. B. & Butler, R. S. (1972) Fractionation with X-rays and neutrons in mice: response of skin and C₃H mammary tumours. *British Journal of Radiology*, 45, 237-249.
Hall, E. J. (1972) Radiation dose rate: a factor of importance in radiobiology and radiotherapy. *Bristish Journal of Radiology*, 4, 81-97.
Holsti, L. R., Rissanen, K. & Spring, E. (1971) Radiobiological determination of the total dose in radiotherapy. *Acta Radiologia* (Therapy Physics Biology) Vol. 10, 289-297
Mallik, A. (1972) The treatment of cervix cancer using high activity 60 Co sources. *British Journal of Radiology*, 45, 257-270.
Liversage, W. D. (1969) A general formula for equating protracted and acute regimes of radiation. *British Journal of Radiology*, 42, 432-440.
Marks, R. C., Agarwal, S. K. & Constable, W. C. (1973) Increased rate of complications as a result of treating only one prescribed field daily. *Radiology*, 107, 615-619.
Norin, T. et al., (1971) Conventional and super-fractionated radiation therapy in Burkitt's Lymphoma. *Acta Radiologica* (Ther), 10, 545-557.
Paterson, R. (1948) *The Treatment of Malignant Disease by Radiotherapy*. London: Arnold.
Pierquin, B. (1970) L'effet differential de l'irradiation continue (ou semi-continue) a faible debit des carcinomes epidermoides. *Journal de Radiologie et l'Electrologie*, 51, 533-536.
Shukovsky, L. J. & Fletcher, G. H. (1973) Time-dose and Tumour Volume Relationships in the irradiation of squamous cell carcinoma of the tonsillar fossa. *Radiology*, 107, 621-626.
Suit, H. D. (1972) Consideration of fractionation schedules for radiation dose. *Radiology*, 105, 151-160.
Suit, H. D., Lindberg, R. & Fletcher, G. H. (1966) Prognostic significance of extent of tumour regression at completion of radiation therapy. *American Journal of Roentgenology Radium Therapyane Nuclear Medicine*, 84, 1100-1107.
Von Essen, C. F. (1963) A spatial model of time-dose-area relationship in radiation therapy. *Radiology*, 81, 881-885.
Wilson, C. S. & Hall, E. J. (1971) On the advisability of treating all fields at each radiotherapy session. *Radiology*, 98, 419-424.
Wiernik, G., Bleehen, N. M., Brindle, et al. (1972) Fifth Progress Report on the British Institute of Radiology fractionation trial of 3F/week of 5F/week treatment of larynx and pharynx. *British Journal of Radiology*, 45, 754-756.

FURTHER READING

Fletcher, G. H. & Shukovsky, L. J. (1975) The Inter-play of radio-curability and tolerance in the irradiation of human cancers. *Journal de Radiologie et l'Electrologie*, 56, 383-400.

15. Developments in Radiotherapy

Several important findings have been reported in the field of radiobiology in the last 10 years that have been relevant to clinical radiotherapy and have influenced developments in clinical practice. Together with an improved understanding of the biology of cancer, progress in these basic sciences has indicated areas in which new techniques should be exploited to improve the results of radiotherapy. Further exploration and development of these experimental approaches may possibly, in the next few years, increase considerably the effectiveness of radiotherapy in the management of many cancers in which only moderate success is achieved at present and also extend its present scope by demonstrating good control rates in cancers now considered unresponsive to radiations normally employed to-day.

Radiation is essentially a 'local' form of treatment and although too many patients still die with uncontrolled local recurrence, a quite separate problem is the treatment of widespread metastatic disease. New techniques, such as cytotoxic chemotherapy, specific and non-specific immune stimulation, and systemic hyperthermia may soon offer the opportunity of eradicating disseminated malignant disease while also enhancing the effects of radiotherapy on the primary cancer. Already evidence of this effect can be found in the improved survival results reported in children with Wilms' tumour, treated by radiotherapy to the abdomen and systemic Actinomycin D and Vincristine. Similar improvements are being produced by supplementary chemotherapy in children with rhabdomyosarcoma, Ewing's tumour and neuroblastoma.

It is now intended to describe new physical, chemical and biological techniques that may increase the effectiveness of radiotherapy in the local control of cancer.

Physical techniques

High LET radiations

High LET radiations that have been proposed for clinical use are fast neutrons, negative pi-mesons and light atomic nuclei such as neon, helium or carbon accelerated in nuclear physics machines. Their biological effects show three important differences from those X-rays and other sparsely ionising radiations (Fig. 15.1). Other biological differences, of less clinical importance, have been described in Chapter 13.

Firstly, it is known that the dose-modifying effect of oxygen is much less following irradiation with high LET than with X or γ-rays. It is for this reason that these new radiations are being evaluated for clinical radiotherapy in the belief that the cure rate of many tumours may be limited by the hypoxic tumour

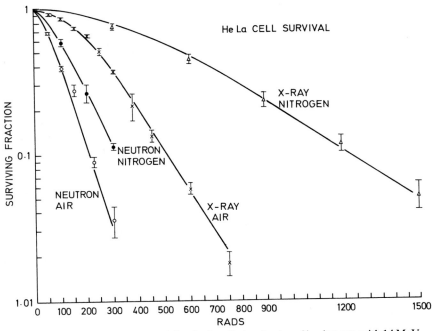

Fig. 15.1 HeLa cell survival curves following irradiation, in air and in nitrogen, with 14 MeV neutrons and 250 kV X-rays.

cells which are relatively resistant to low LET radiations. Another important difference is that the capacity of cells to shed sublethal radiation damage is much reduced, or absent, after exposure to high LET radiations. This is manifest in cell survival curves by a reduction in the size of the shoulder (D_q value) and if this change is greater for some types of cancer cells than for cells in normal tissues then the therapeutic ratio for these particular tumours would be increased. The damage to the cancer cells would as a result be more accumulative over a fractionated course of treatment than to the surrounding normal tissues that had a better capacity to shed sublethal damage.

Thirdly, all mammalian cells show a greater sensitivity (lower D_o values) to high LET radiations than to X-rays. The differences observed in D_o or D_q values determines that the biological effectiveness of fast neutrons is greater than X-rays. It must be understood also that because of the differences in the shoulder regions, that is the capacity to recover from sublethal damage, after irradiation with fast neutrons or X-rays, the RBE will change with size of dose per fraction. The lower the dose per fraction the greater is the RBE (Fig. 15.2). This change level of RBE with the size of dose fraction may result in unwary clinicians falling into a 'fractionation trap'. The relatively low value of RBE obtained for high doses given as a single fraction will increase significantly when smaller dose fractions are given in a fractionated course of treatment. This effect was responsible for many patients being given too high treatment doses in the first clinical trial of fast neutron therapy. It has also to be noted that

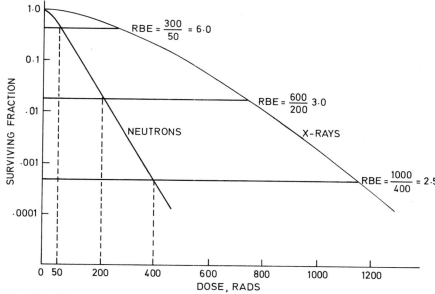

Fig. 15.2 Schematic cell survival curves for fast neutrons and X-rays that illustrate the increase in RBE with decreasing dose fraction. This effect is due to the lack of 'shoulder' on the survival curve associated with high LET radiation.

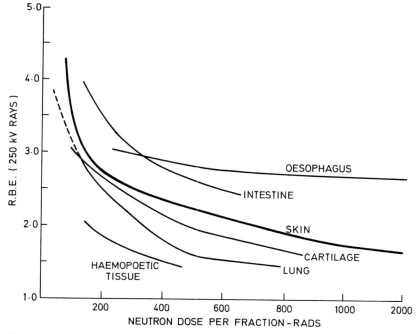

Fig. 15.3 The relationship of RBE to the dose per fraction of 7 MeV neutrons for a number of mammalian tissues. (From Hornsey & Field, 1975)

the RBE values as a function of size of dose fraction varies from tissue to tissue. (Fig. 15.3). (Hornsey and Field, 1975).

One of the most promising new techniques in radiotherapy is the use of fast neutron beams. *Fast neutrons* can be produced by accelerating deuterons on to a tritium target resulting in a fusion reaction from which mono-energetic 14.7 MeV neutrons are ejected. This reaction may be produced in specially designed D–T neutron generators. Alternatively, neutrons may be produced in a cyclotron, usually by the bombardment of a beryllium target with deuterons of 16 to 50 MeV. In this reaction the neutrons are stripped from the deuterons in the beryllium target and a spectrum of neutron energies is produced, the mean energy of which is just under half that of the deuteron beam. Because they are uncharged particles the depth dose curves of fast neutrons are very similar to those of high energy photons.

The interactions with tissues are quite different from the attenuation of X and X rays and have already been described in Chapter 2. These differences mean that the relative absorption of energy in the tissues is not the same as with X-rays (Fig. 15.4). At a rough approximation the absorption of neutrons is

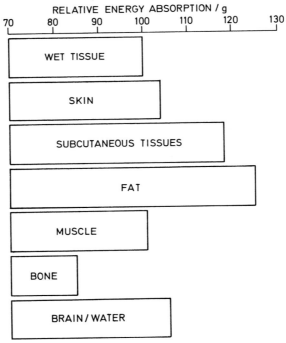

Fig. 15.4 The relative absorption of fast neutrons in tissues.

proportional to the concentration of hydrogen in the tissues. Fat, which is rich in hydrogenous material, will absorb much more energy from neutrons than from X-rays. By contrast, bone which contains a high proportion of the higher atomic number elements calcium and phosphorus will absorb proportionately less neutron radiation than low LET radiation. Accordingly neutrons will have a relatively sparing effect on bone while subcutaneous tissues may be expected to absorb some 15 to 20 per cent more energy from neutron radiation than from X-rays.

Fast neutrons have been shown to be used clinically with safety in a series of investigations conducted at the Hammersmith Hospital, London, using a neutron beam with a mean energy of 7.5 MeV. A randomized clinical trial of fast neutron therapy in patients with advanced head and neck cancers has been reported (Catterall *et al.*, 1975). It has shown (Table 15.1) significant advan-

Table 15.1 Actuarial assessment of short-term control rates for 'head and neck' cancers treated with fast neutrons or megavoltage X-rays. (Hammersmith Trial). (From Catterall, Bewley & Sutherland, 1975).

	Neutrons	Photons
Number of Patients	52	50
Complete Regression	37 (71%)	16 (32%)
Later Recurrence	0	9
Local Control	37 (71%)	7 (14%)

tage in the rate of local tumour control in the group of patients treated by fast neutrons. The morbidity was apparently little different in the patients treated with photons than in those given fast neutron therapy, but it is important that a longer period of observation be maintained before one can be certain about the real improvement in effectiveness of fast neutrons. It has been suggested too that some cancers which do not respond very well to X-ray therapy, such as carcinomas of the gastro-intestinal tract and salirary glands as well as sareomas may well be controlled much more effectively with fast neutrons. The results of these early trials are promising and further trials are now justified.

The results of preliminary clinical trials have also been reported from the M D Anderson Hospital and Tumour Institute using neutrons of mean energy of 6 MeV and 22 MeV from a large variable energy cyclotron. A miscellaneous group of tumours was treated and much of their investigation (Hussey *et al.*, 1974) was concerned with evaluation of dose-related effects on normal tissues to determine tolerance dose schedules. It was found that treatment over seven weeks using two fractions per week caused more late necrosis than would have been expected from the early reactions, both for neutrons and for photons. However, some encouraging clinical results in the number of short-term regressions (up to 12 months) were observed in patients with advanced cancer of the head and neck region and also in others with breast cancers.

Sources of neutron radiation suitable for intra-cavitary and interstitial use are also available in the form of Californium-252, or a mixture of curium-242 and beryllium. There are great technical difficulties in measuring OER at very low dose-rates, but *in vitro* measurements that have been made indicate that the OER for Californium-252 may be significantly lower than that obtained with comparable dose rates using radium. The gain factor compared to radium is about 1.5 (Hall & Rossi, 1975). If this gain factor is confirmed then these neutron sources may have some advantage, at dose rates of about 10 rads per hour, over the use of low dose-rate implant techniques using radium or other γemitting radio-nuclides. There would be a smaller therapeutic gain when dose-rates associated with intra-cavitary radium and caesium techniques for cancer of the cervix are used and further evidence is required to determine whether this would be clinically significant. In view of the very high RBE values for small doses and low dose-rates of neutrons, the radiation hazards are considerable. The difficulties this introduces in the intra-cavitary and interstitial application of these neutron sources requires that the possible clinical benefits are clearly established by experimental studies before their clinical use should be generally adopted.

Negative π-mesons are sub-atomic particles with a mass 276 times that of an electron, but with the same negative charge. They are produced by accelerating extremely high energy protons (500 to 750 MeV) in a synchrocyclotron on to a graphite or lead target. Negative π-mesons are interesting because the deposition of energy occurs mainly in the Bragg ionisation peak where the pions slow down and are captured by nuclei present in tissues. Pions produce densely ionising particles when they are attenuated in tissues as a result of nuclear disintegrations releasing alpha-particles, neutrons and protons following their capture in the nuclei of carbon, oxygen and nitrogen. The entrance, or plateau, region of the depth-dose curve is produced by the much less densely ionising fast particles and so in this region the RBE of the pion beam is much lower (Fig. 15.5). The RBE values in the plateau region are about 1.0, but in the peak

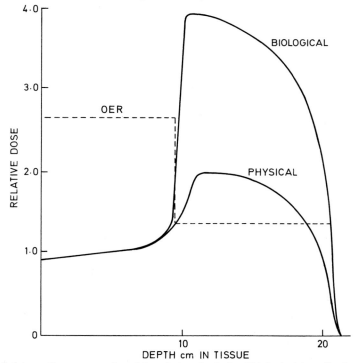

Fig. 15.5 Schematic representation of the relative physical and biological dose distribution of negative π-mesons in tissues. The change in value of the OER is also of prime importance.

region the RBE depends on the dose and the biological system being examined. A range of 'peak' RBE values from 1.6 to 3.0 has been demonstrated. Negative π-meson beams therefore present the practical possibility of delivery of radiation with a reasonably low OER ($\sim$ 1.8) to a well circumscribed volume of tissue. (Bond, 1971).

Clinical facilities are being established for π-meson therapy in the United States at Los Alamos and Stanford, in Canada at Vancouver, and in Europe at Zurich and possibly at Harwell.

Light atomic nuclei of carbon, helium, neon, argon and other elements have also been the subject of experimental evaluation with a view to clinical application (Hall, 1973). These highly energetic charged particles have physical and biological properties somewhat similar to negative π-mesons. (Table 15.2).

Table 15.2 Comparative Advantages of Radiation Beams for Cancer Therapy

Radiations	Biological & Physical Criteria	
	Low OER	Defined Dose Distribution
X and Gamma	*	*
Protons	*	***
Fast Neutrons	**	*
Helium Ions	*	***
Negative π-mesons	*	**
Heavy Nuclei	**	*

Helium ions produce good dose distributions, similar to protons, and the OER in the peak region may be slightly lower, but this has to be confirmed. High energy heavy nuclei offer theoretical improvements in both the biological and physical characteristics of the treatment beam, but there is at present inadequate experimental data available. It is to be expected that the OER may be greatly reduced in the peak region of the depth-dose curve and for heavy nuclei of, for example, argon the OER may be near unity in the last few centimetres of its range. The depth dose distribution, however, may not be as circumscribed as obtained with protons that have a much wider Bragg peak. In the case of heavy ions the addition of multiple peaks to produce an adequate high-dose volume may result in a smaller peak to plateau ratio. Some clinical studies have been conducted on patients with mycosis fungoides and other skin cancers (D'Angio *et al.*, 1974). It is not yet possible, however, to make any definitive assessment of the eventual role these beams may have in clinical practice.

Beams of high LET radiations are difficult and expensive to obtain and their shielding and collimation are difficult for clinical applications. It may be of practical importance that it has been shown that high LET radiations and photons given within a short time interact in an interesting manner. As a result the OER of the combined beams remains low (1.6) when half the radiation is of low LET quality and in the range of dose fractions of mixed irradiations the R.B.E. has a predictable linear relationship. It may be expedient to plan a regime of radiotherapy which combines both high and low LET radiations given within an hour. This would allow better dose distribution in the tumour and better shielding of normal tissues than may normally be possible using high LET radiations alone. And yet with a 50 per cent contribution of X or γ-rays or electrons a high gain factor might still be obtained in respect of the hypoxic cancer cells.

Hyperthermia and radiotherapy

It is over 100 years since artificial elevation of the body temperature was first used to attempt to cure cancer, and over forty years since it was first demonstrated clinically that the combination of heat and X-irradiation could enhance the expected response of cancer to radiation alone. Recently there has been a re-birth of interest in the effects of hyperthermia on cancer and the demonstration that it is a powerful synergist of radiation effects.

The mechanism of the action of hyperthermia on normal and cancer cells is still poorly understood, but it has been shown that some cancer cells are more sensitive to heat treatment than normal cells. This selective effect is much more marked above about 42°C. The mode of action is certainly different from that of irradiation and the effects of these differences may have important implications for radiotherapy. Cells in the late S-phase which are the most resistant to X-rays are found to be the most sensitive to the effects of heating. The combination of heating and X-radiation therefore produces a reduction in the differential radio-sensitivity of cells, throughout the cell cycle potentiating the effects of radiation on asynchronous populations. It has also been shown that whereas there is a direct relationship between chromosome aberrations and the cell killing throughout the cell cycle following irradiation, this correlation applies only to S-phase cells following hyperthermia.

The quantitative cellular effects of heating depend critically on the temperature to which they are elevated and the period for which they are kept at that temperature. In general terms there is a complementary relationship between temperature and time so that the proportion of cells sterilized may be correlated with the total energy absorbed. If irradiated with simultaneous hyperthermia the radiation cell-survival curve shows a greatly reduced shoulder region and the slope of the exponential portion of the curve is much steeper. It is interesting that although mammalian cells can repair sub-lethal damage within a few hours after heating, recovery from sub-lethal radiation damage is reduced in heated cells.

It is also of importance that hypoxic cells are more sensitive to hyperthermia than well-oxygenated cells and so the combination of hyperthermia and radiation may be yet another method of helping to overcome the radio-resistance of hypoxic tumour cells. Robinson, Wizenberg & McCreadie (1974) have shown a reduction in O.E.R. from 3.0 at 37°C to 1.4 at 43°C. (Fig. 15.6). They have also shown that the differential lethal effect of heating on hypoxic cells increased more rapidly than that for well-oxygenated cells, as the temperature increased.

Techniques of inducing either systemic hyperthermia or localized heating for example by microwave radiation are currently being evaluated in a few cancer

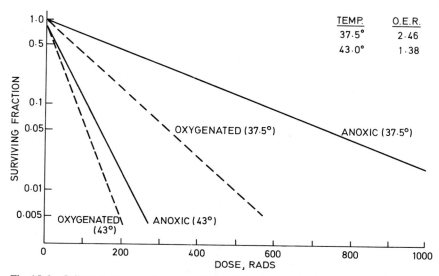

Fig. 15.6 Cell survival curves for mammalian cells irradiated with X-rays, in air and in nitrogen, at normal and elevated temperature. There is a marked fall in OER from 2.46 to 1.38. (From Robinson *et al.*, 1974)

centres. Basic radiobiological research continues to investigate the subject further as there is the possibility that the combination of hyperthermia and radiotherapy may prove to be of clinical benefit. A great deal more fundamental research is required before properly planned clinical trials can be undertaken.

Chemical techniques

Combination of drugs and radiation

In the treatment of many tumours the probability of local control is much lower than one would hope and this is particularly true when the growth is large; at other times the histology of the tumour is unfavourable and indicates a poor response to radiation alone. Attempts have been made to increase the radiation response by administering radio-sensitizing agents. With most agents the normal tissues within the field of irradiation are equally sensitized and so no real improvement in the differential response between the cancer and normal tissues is obtained. This field of research continues to be explored and some extremely interesting developments may now be reported..

It should first be explained that drugs and radiation may be usefully combined in the treatment of cancer for two distinct and separate reasons. Drugs may be used in a *complementary* role to increase the local effects of radiation within the treated volume. Another role that is becoming increasingly important is the use of *supplementary* chemotherapy. By this is meant the administration of cancer chemotherapeutic agents primarily to destroy disseminated cancer cells outside the volume of irradiation. All of these drugs may also influ-ence, to some extent, the local tumour response, but their administration may often be started after the course of radiotherapy has been completed. They will normally have no additive or synergistic effect on the radiation response of the primary cancer and are not 'radio-sensitizers'.

The drug-radiation interaction

The effects of radiation may be increased by chemotherapeutic agents in three ways (Fig. 15.7). Firstly their actions may simply be *additive*, the total biological effect being no greater than the lethal effect on cells of both the dose

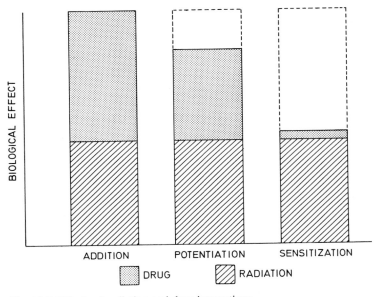

Fig. 15.7 Effects of radiation and drug interactions.

of the drug and radiation independently. Secondly, the radiation effect may be *potentiated* by the drug, so that their combined effect is greater than that of the individual treatments. And thirdly, there may be the true radiation *sensitiza-tion* where the drug itself has little or no lethal effect on the cells, but so influ-ences their response to radiation that the combined biological effect is augmented. Potentiating agents and radio-sensitizers may alter the radiation effect in two ways, either by increasing the intrinsic sensitivity (D_o value) of the cells, or by reducing their capacity to recover from sub-lethal damage (mea-sured by the D_q value).

Additive chemotherapy

It is doubtful if this form of combination therapy offers any real advantage in cancer control. The alkylating agents have been used in this way without any evidence that their addition has increased the local tumour control even in the management of lympho-reticular diseases. It is often considered advisable to add an alkylating agent to the X-ray treatment of extra-dural deposits of Hodgkin's Disease, when there may be doubt that the tolerance dose of radiation to the spinal cord will be sufficient to eradicate the local tumour. It is in situations where the normal tissue response to the drug may be significantly different from that following radiation that benefit of additive chemotherapy may be helpful. As yet no drug which is suitable for this type of combined treatment has yet been evaluated.

Potentiating agents

This is the most common group of drugs that have been used to complement the effects of radiation. All have cancerocidal action in their own right and in these groups are included 5-fluoro-uracil, methotrexate, actinomycin D and bleomycin. 5-fluoro-uracil is one of the halogenated pyrimidines and blocks the enzyme thymidylate synthetase. Unlike other halogenated pyrimidines it is probably only a weak radiation sensitizer and its effect mainly additive. It has been shown in experimental systems to produce a synergistic effect with radiation at certain times in relation to the exposure to radiation, while at other times it was only additive. A number of clinical trials have been conducted and they have tended to demonstrate that there is probably little advantage in combined therapy with 5-fluoro-uracil.

Methotrexate is another anti-metabolite, a folic acid antagonist that alters the response of cells to X-rays as well as itself being cytotoxic. After treatment with methotrexate the shoulder of the survival curve is greatly reduced and to some extent the intrinsic radio-sensitivity of the cells is increased. In addition, it has also been shown that the intrinsic radio-sensitivity of hypoxic cells is increased much more than well-oxygenated cells after exposure to methotrexate.

Again several clinical trials have attempted to demonstrate benefit from combined treatment with methotrexate and radiation, but no convincing results have been published.

Actinomycin D is a cytotoxic antibiotic which depresses DNA dependent RNA synthesis. It has been shown to increase the radio-sensitivity of cells, (measured by decrease in D_o value) and will, to a lesser extent, diminish the repair capacity, (D_q value). It is, however, true that actinomycin D has not been shown to be of added benefit in curing irradiated tumours. It has been shown to be extremely effective in controlling small metastatic deposits from Wilms' tumour and from Ewing's tumours in areas which have not been irradiated and in this supplementary role it has made an impressive improvement in survival rates.

Bleomycin is another antibiotic which is particularly interesting for it has an affinity for keratinizing squamous epithelium and is also selectively concentrated in squamous cell carcinomas. There is some evidence that bleomycin does increase the radio-sensitivity of mammalian cells and may be an important potentiating agent.

Doubts exist about the mode of action of these drugs and their modification of radiation response. The *in vitro* experimental studies have been undertaken using mixed cell populations. These agents are cytotoxic mainly to cells in the S-phase, which are the most resistant to radiation, and so the apparent increased radio-sensitivity of the total population could perhaps be explained by the selective removal of a part of the resistant population. Certainly further basic research is needed before clinical trials can be properly planned for their evaluation.

Radio-sensitizing agents

Oxygen is one of the most powerful radiation sensitizers and the importance of the oxygen effect in radiobiology has been stressed in earlier chapters. Many methods of improving the oxygenation of tumours have been tried, in order to reduce the radio-resistance of hypoxic cells in tumours. (Duncan, 1973)

The most widely used technique has been the administration of oxygen at high pressure during irradiation which saturates the plasma and tissue fluids with oxygen. The amount of oxygen available is greatly increased in this way by the increased diffusion gradient from the capillaries to the hypoxic tumour cells. Normally oxygen at three atmospheres of pressure is employed which will allow diffusion to a distance of about 1000 μm instead of the usual 100 to 200 μm.

Unfortunately there are two major problems concerning the clinical use of hyperbaric oxygen techniques in radiotherapy. The first problem is the uncertainty that the oxygen dissolved at high pressure in the plasma does actually reach the hypoxic cells in tumours. Any obstruction of the vascular tree will greatly reduce the oxygen available to the related tissues and vaso-construction is a known response to high oxygen tension (Fig. 15.8). The second disadvantage is the possibility that normal tissues may be sensitized to some extent and although less than the tumour, no great improvement in therapeutic ratio may be obtained.

There are also a number of physical and clinical dangers which are related to the administration of hyperbaric oxygen. The major hazard is fire and great care has to be taken in its prevention; explosive decompression is also a possibility although they are very rare hazards indeed. A clinical difficulty sometimes met is the claustrophobia experienced by about 10 per cent of the patients. Oxygen convusions can occur and certain chest complications and ear troubles due to pressurisation.

The results of the first random controlled clinical trial was published from Cardiff (Henk *et al.*, 1970). In this trial 213 patients with various cancers of the head and neck region were treated either in air or hyperbaric oxygen at three atmospheres. Patients were given the same dose (3500–4000 rads in 10 fractions over three weeks) both in air and in oxygen. The local control rates of tumours were significantly better in the group of patients irradiated in hyperbaric oxygen (Table 15.3). However, the morbidity in patients treated in hyperbaric oxygen was greater than in air, and so it is certain if the therapeutic ratio has been increased. Would a slightly higher dose of X-rays given in air

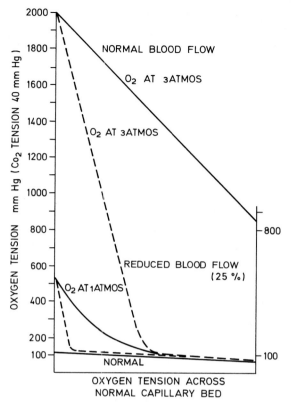

Fig. 15.8 The considerable improvement in capillary oxygen tension is shown when oxygen at 3 atmospheres pressure is breathed. If there is any obstruction to the blood flow this improvement is largely lost. Stagnant anoxia in tumours may negate the possible advantages of irradiating tumours in hyperbaric oxygen.

have given just as good a result? The results of further trials have to be awaited completely to resolve this question, but the evidence is suggestive that in some cancers X-ray treatment in hyperbaric oxygen provides a real advantage over treatment in air.

An important radiobiological consideration which could influence clinical results is the size of dose of X-rays given at each treatment in hyperbaric

Table 15.3 Local control rates at two years following X-ray treatment in air or in hyperbaric oxygen (Cardiff Trial). (From Henk *et al.*, 1970).

| Tumour Site | Treated in Air | | Treated in HPO | | |
	Number	Local Control %	Number	Local Control %	Statistical significance
Nasal & Oral	15	24.8	18	51.8	p = 0.01
Laryngeal	9	45.3	10	77.9	p = 0.05
Laryngo-pharyngeal	10	33.6	8	51.1	—
All sites	34	35.1	36	54.1	p = 0.05

oxygen. Figure 15.9 illustrates the cell survival curve for a mixed population of well-oxygenated and hypoxic cells. In the initial part of the survival curve the response is predominantly determined by the proportion of well-oxygenated cells and so with small fractionated doses of 200–300 rads, no differential effect

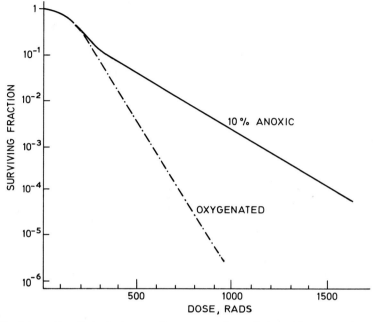

Fig. 15.9   Schematic cell survival curve of a tumour cell population with a proportion of anoxic cells that produce the radio-resistant 'tail' at higher dose levels. The initial slope is predominantly determined by the response of well-oxygenated cells and so a biphasic curve is characteristically obtained.

might be expected unless there was some change in the shape of this cell survival curve, (such as occurs due to re-oxygenation). It is only when higher dose fractions are given in hyperbaric oxygen that a reduction of the oxygen effect might be demonstrated in some cancers. The original clinical studies of X-ray therapy in hyperbaric oxygen utilized such larger doses and several clinical trials have now been commenced which take this effect into account.

The halogenated pyrimidines, 5-chloro, bromo and iodo-deoxyuridine closely resemble the structure of thymidine, an essential DNA precursor, in that its methyl group is replaced by the halogen (Fig. 15.10). By competition these compounds may replace thymidine in the DNA chain and the cells incorporating them will show increased sensitivity (lower D_o values) and diminished capacity of recovery from sub-lethal radiation damage (lower D_q values). This effect is not fully understood, but is probably due to an increase in the number of radio-chemical lesions produced in DNA as a result of 5-BUdR incorporation, rather than a weakening of the DNA chain structure following thymidine replacement.

An important practical disadvantage to the use of the halogenated

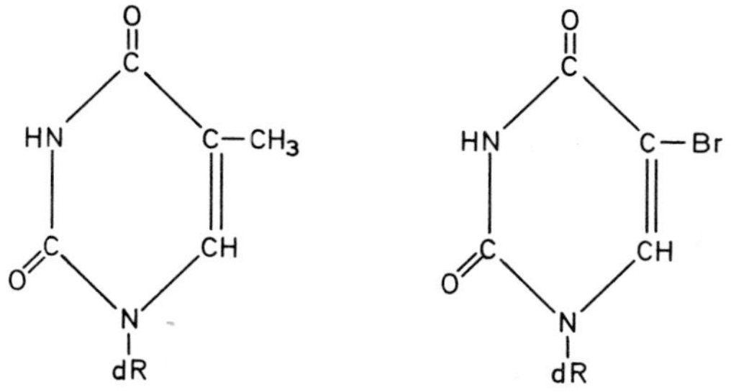

Fig. 15.10 Similarity of 5-Bromodeoxyuridine to Thymidine with which it competes for incorporation into DNA

pyrimidines is that they are rapidly degraded in the liver and to be effective they have to be infused intra-arterially before each exposure to X-rays. Another drawback is that only cells synthesising DNA will incorporate these drugs, although the uptake can be increased to some extent by inhibiting thymidylic acid synthesis by giving methotrexate or 5-fluoro-uracil. The 5-BUdR compound has been shown to be the least toxic of the halogenated thymidine analogues, but they all produce toxic side effects on the normal proliferating cell systems such as bone marrow, intestinal epithelium and the hair follicles. It is likely too that a good differential sensitization of tumour cells compared to normal tissues in the radiation field will only be obtained if the cell cycle time of the tumour is *much shorter* than that of the normal tissues. This is seldom the case in human tumours, with perhaps the only exception being that of gliomas of the cerebral hemispheres. An uncontrolled clinical trial has been undertaken (Hoshino *et al.*, 1970) which provided encouraging results, but we must await conclusive evaluation of this method of treatment by carefully designed and properly controlled trials.

Hypoxic cell radio-sensitizers

Considerable progress has been made in the development and study of chemical substances that will selectively sensitize hypoxic cells (Adams, 1973). If these substances can reach the hypoxic cancer cells that may form radio-resistant cords around the capillaries in a tumour, X-ray therapy would be much more effective in respect of these hypoxic cells by their prior administration, while the sensitivity of normal and well-oxygenated cells would remain unchanged.

It has already been mentioned that methotrexate has been shown to be a sensitizer of hypoxic cells. We wish here to draw attention to a group of substances that mimic the effect of oxygen. The early work in this field was disappointing in that the nitroxyl group of compounds (e.g. TAN tri-acetoneamine-N-oxyl), which sensitized bacteria, failed to be effective at non-toxic concentrations with mammalian cells. Interest presently is focussed on a large class of compounds characterised by a strong electron affinity. They include substances known as glyoxals, quinones, aceto-phenones, nitro-furans and nitro-imidazoles. These compounds, like oxygen, sensitize hypoxic cells in all phases of the cell cycle. One of the quinone group is the substance menadione, which is the active form of Synkavit which has been carefully investigated without any convincing clinical benefit being demonstrated.

Many of the nitro-furans are now used in other clinical applications because of their anti-bacterial properties. *In vitro* studies have shown that mammalian cells under hypoxic conditions are sensitized by both nitro-furantoin and nitro-furazone. Further investigations are being conducted into the relative effectiveness of other nitro-furans which may require less high concentrations, but their use in animals is limited because of their too rapid metabolism.

Perhaps the most exciting group of all is the nitro-imadazoles of which the commonly used drug metranidazole (Flagyl) is a member. It has been shown that this drug in high concentration is a powerful radio-sensitizer of hypoxic cells only, and the use of these radio-sensitizing drugs is being further explored. A drug similar to Flagyl, (1-(2 nitro-1-imidazolyl) 3-methoxy 2 propanol) has been shown to be a particularly powerful radio-sensitizer for mamallian cells. In many different types of experimental tumours in mice this drug has increased the response to X-rays by a factor of 2 in some cases. Clinical trials are now in progress and if this degree of sensitization can be obtained in human tumours, dramatic improvements in local control rates of many cancers may be expected. (Adams *et al.*, 1976).

Biological techniques

Fractionation

Most courses of radiotherapy give small daily fractions, the total dose having been established by clinical experience over many years. Because of the several factors that influence biological response, it must be clear that this regime is unlikely to be optimum for all tumours that may arise in many different tissues. Rational design of the size and spacing of dose fractions will only be possible when data are known about the important biological parameters described in Chapter 12, intrinsic radio-sensitivity, recovery, repopulation re-oxygenation and redistribution. The application of different fractionation schedules has been discussed in Chapter 14.

One unconventional method is the split-course regime in which two courses of radiotherapy are separated by a rest interval which allows repopulation of normal tissues and in some situations re-oxygenation of the tumour. Most commonly daily treatments have been given for two or three weeks before an interval of about two weeks, after which daily treatments are resumed for a further two or three weeks. Results of trials conducted so far have not indicated any real improvement in therapeutic ratio, for although immediate radiation reactions are much less severe the late reactions are not reduced.

Recently results of the British Institute of Radiology trial of 3 and 5 fractions each week in the treatment of cancer of the larynx and pharynx showed no difference in survival. It would seem that minor changes in fractionation of this type are unlikely to show any major therapeutic advantage, which will have to await detailed information about cell kinetics. It should be remembered that reduced fractionation, if providing no worse clinical results, may be of great economic advantage which should not be ignored.

Kinetic analysis

It will be recognized from earlier chapters that information about the number of clonogenic tumour cells and the proliferative state would be extremely helpful in the planning of fractionated radiotherapy (Brown, 1975).

The most important data to determine is the rate of proliferation of the tumour cell population and that of the normal tissues irradiated in the high-dose volume. Just as with other parameters affecting the response to radiation, there is no evidence that there is any consistent difference in the proliferation rates of the stem cells in tumours and normal tissues. However, detailed information on the range of cell cycle times of tumours is just beginning to be

Table 15.4 Cell Kinetics of Human Solid Tumours. (After Terz *et al.*, 1971)

Tumour Type	Cell Cycle Time (hrs)	DNA Synthetic Period (hrs)	Growth Fraction %	Cell Loss %
Carcinoma Cervix	15	10	50	—
Carcinoma Colon	26	14	45	40
Epidermoid Carcinoma	38	12	24	90
Carcinoma Breast	42	21	43	—
Malignant Melanoma	72	21	25	70
Basal Cell Carcinoma	72	19	30	95

known (Table 15.4) and perhaps differences which could be explored clinically may be discovered (Tubiana, 1971).

Information of this kind is necessary in order to attempt to enhance the radiation response by cell synchronisation using chemotherapeutic agents. One immediate, and perhaps insurmountable, difficulty is that if there is a wide range of intermitotic intervals in the proliferating tumour cells it may be impossible to bring a significant proportion of tumour cells into the sensitive phase of the cell cycle at the time of irradiation. In addition many tumours may contain a high proportion of cells in the resting phase that cannot be manipulated in this way and remain relatively resistant. A great deal more information is required about the proliferation rates and the proportions of non-cycling cells in both tumours and normal tissues before such techniques could be used to influence favourably the therapeutic ratio.

It should be added that synchronisation of normal cells alone may be possible so that irradiation could be undertaken at a relatively resistant phase compared to tumour and techniques of this kind are being explored.

Cell synchronisation

The large difference in radio-sensitivity that may be measured throughout the cell cycle has been already described (Ch. 5). Attempts have been made to synchronise the tumour cells so that they may be irradiated subsequently at the most sensitive phase of the cycle. Clinical applications of these techniques that may be demonstrated *in vivo* have all failed to show any benefit. In addition the *in vitro* techniques (by culture, H3 thymidine labelling and auto-radiography) required to assess synchronisation take so long that they cannot provide timely information for clinical use. Another problem remains that normal tissues will also be synchronised to some degree and this may be disadvantageous. It also appears to be extremely difficult to maintain even partial synchronomy throughout a course of radiotherapy.

Methotrexate and bleomycin, which inhibits progression at the early and late S-phase respectively, have been used in this way and partial synchronisation has been demonstrated in human tumours. Similarly, hydroxyurea, which holds up cells at the beginning of S-phase and which may be used to effect synchrony, *in vitro*, has been tried in clinical trials of head and neck cancer. It is doubtful, however, if any synchronisation occurred and hydroxyurea is considered too toxic at the concentrations required for effective control of cell cycle progression.

Radiation itself will produce some synchronisation by selectively killing these cells in the most sensitive phase, mitosis, and by usually blocking the progression of cells in the post-synthetic (G_2) phase of the cycle. The concept of cell modulated dose fractionation has been suggested, but too many imponderables exist to permit its application. A large conditioning dose would be given to produce some synchronisation, and some recruitment of resting cells into the cell cycle and the next dose of radiation would be given when the cells next reached the most sensitive phase. Sufficient information is not available to produce, or to measure, in human tumours and their related normal tissues, the proportion of cells in each phase and the timing of progression following attempts at synchronization. Inappropriate timing of fractionation schedules based on inadequate or inaccurate data could of course lead to deterioration in local tumour control and an increase in morbidity.

Conclusion

It has to be acknowledged that, although *many* cancers are eradicated by modern radiotherapy techniques, too often there is failure to control the primary tumour. The dose of radiation may be limited by either local or systemic tolerance. These experimental clinical techniques, many of which have not yet been properly evaluated, aim to enhance the therapeutic ratio by increasing local tumour control without exceeding normal tissue tolerance. There is no doubt that we require much more information about the basic action of many of these new techniques and also about their possible interaction for it may prove advantageous to use a combination of these different techniques for optimum effect. Already we have seen in clinical practice great improvements as a result of judicious combination of chemotherapy with radiotherapy and of surgery and radiotherapy. There can, however, be no doubt that further exploitation of many of these new radiotherapy techniques, based on radiobiological principles, will progressively contribute to the continuing improvement of cure rates in cancer therapy.

REFERENCES

Adams, G. E. (1973) Chemical radiosensitization of hypoxic cells. *British Medical Bulletin,* **29,** 48–53.
Adams, G. E., Fowler, J. F., Dische, S. & Thomlinson, R. H. (1976) Hypoxic cell sensitization in radiotherapy. *Lancet,* **i.** 186–188.
Bond, V. P. (1971) Negative pions—Their possible use in radiotherapy. *American Journal of Roentgenology, Radium Therapy and Nuclear Medicine,* **11,** 9–26.
Brown, J. M. (1975) Exploitation of kinetic differences between normal and malignant tissue. *Radiobiology,* **114,** 189–198.
Catterall, Mary, Sutherland, I, Bewley, D. K. (1975) First results of a randomised trial of fast neutrons compared to X-rays in treatment of advanced tumours of the head and neck. Report to the Medical Research Council. *British Medical Journal,* **2,** 653–655.
D'Angio, G. J., Aceto, H., Nisce, L. Z., Kim, J. H., Jolly, R., Buckle, D. (1974) Preliminary clinical observations after extended Bragg peak helium irradiation. *Cancer,* **34,** 6–11.
Duncan, W. (1973) The exploitation of the oxygen effect in clinical practice. *British Medical Bulletin,* **29,** 33–38.
Hall, E. J. (1973) Radiobiology of heavy particle radiation therapy: cellular studies. *Radiology,* **108,** 119–129.
Hall, E. J., & Rossi, H. (1970) The potential of Californium-252 in radiotherapy. *British Journal of Radiology,* **48,** 777–790.
Henk, J. M., Kunkler, P. B., Shah, N. K., Smith, C. W., Sutherland, W. H., & Wassif, S. B. (1970) Hyperbaric oxygen in radiotherapy of head and neck carcinoma. *Clinical Radiology,* **21,** 223–231.
Hornsey, S., & Field, S. G. (1975) On the differences in RBE observed with different tissues and factors governing cell sensitivity, which these differences imply. *Symposium Fast Neutron Dosimetry* I.A.E.A. Vienna.
Hoshino, T., Nagai, M., Sato, F., Sano, K., & Watari, T. (1970) *In Radiation Protection and Sensitization,* pp. 491–497. Edited by H. Moroson and M. Quintiliani. London: Taylor & Francis.
Hussey, D. H., Fletcher, G. H. & Caderao, U. B. (1974) Experience with fast neutron therapy using the Texas A & M variable energy cyclotron. *Cancer,* **34,** 67–77.
Robinson, J. E., Wizenberg, M. J. & McCready, W. A. (1974) Radiation and hyperthermal response of normal tissue. *Radiology,* **113,** 195–198.
Terz, J. J., Curcitchet, H. P. & Lawrence, W. Jr. (1971) Analysis of the cell kinetics of human solid tumours. *Cancer,* **28,** 1100–1110.
Tubiana, M. (1971) The kinetics of tumour cell proliferation and radiotherapy. *British Journal of Radiology,* **44,** 325–336.

FURTHER READING

Biological basis of radiotherapy. (1973) *British Medical Bulletin,* **29,** No. 1.

INDEX